Directed Algebraic Topology and Concurrency

Lisbeth Fajstrup · Eric Goubault
Emmanuel Haucourt · Samuel Mimram
Martin Raussen

Directed Algebraic Topology and Concurrency

 Springer

Lisbeth Fajstrup
Department of Mathematical Sciences
Aalborg University
Aalborg
Denmark

Eric Goubault
École Polytechnique
Palaiseau
France

Emmanuel Haucourt
École Polytechnique
Palaiseau
France

Samuel Mimram
École Polytechnique
Palaiseau
France

Martin Raussen
Department of Mathematical Sciences
Aalborg University
Aalborg
Denmark

ISBN 978-3-319-79217-0   ISBN 978-3-319-15398-8   (eBook)
DOI 10.1007/978-3-319-15398-8

Printed on acid-free paper

This Springer imprint is published by SpringerNature
The registered company is Springer International Publishing AG Switzerland

Foreword

Computer science has a rugged tradition of appropriating concepts and mechanisms from classical mathematics, concepts once proudly considered pure and useless, and adapting those concepts to describe and analyze phenomena that once could not have been imagined. Today, the most abstract discoveries of number theory form the foundations of modern cryptography and finance, graph theory lies at the heart of communication networks, social networks, and search structures, and advanced probability has myriad of everyday applications.

This book, too, presents novel adaptations and applications of basic concepts from combinatorial and algebraic topology: in this case, paths and their homotopies. The intuition is appealing: one can visualize an execution of a concurrent, two-process program as a path winding its way through a planar region, where progress by one process nudges the arrow along the x-axis, and progress by the other along the y-axis. Certain regions of the plane are forbidden: they correspond to zones of mutual exclusion. Two executions are considered equivalent if one can be continuously deformed to the other without crossing a forbidden zone, and the different ways in which paths can snake around these zones define the different ways synchronization mechanisms can shape computation.

These paths, like time itself, cannot run backwards, so the classic theory of paths and their homotopies must be adapted to incorporate a relentless sense of direction, yielding a new theory of directed topology. This theory leads to novel formulations of program semantics, new algorithmic techniques, and a rich description in the language of category theory, itself another area with few prior concrete applications. In the tradition of theoretical computer science, this book takes classic mathematical ideas, shapes them to new and unforeseen applications, and from these ingredients produces a new mathematics.

Providence   Maurice Herlihy
June 2015

Preface

Fascinating links between the semantics of concurrent programs and algebraic topology have been discovered and developed since the 1990s, motivated by the hope that each field could enrich the other, by providing new tools and applications. Soon enough, it turned out that this interaction had to evolve into something richer than a simple dictionary: topological spaces were not exactly the right notion (they must be refined in order to incorporate the direction of time). The algorithms for verifying concurrent programs resulting from topological semantics were not easy to invent while achieving reasonable complexity. Today, we think that enough material has been understood to justify a book on the topic. We felt the urge for a coherent, exhaustive, and yet introductory presentation of the subject, so that it can gain a larger audience and constitute a panorama of the current knowledge upon which future developments will be built.

The topic makes it natural to address both computer scientists and mathematicians. We have done our best to write the book with this mixed audience in mind. Except for the last chapters, we have tried to require few prerequisites, while keeping a reasonable size for the text: only a general knowledge of semantics of programming languages is required, as well as basic notions of (algebraic) topology and category theory.

We thank Thibault Balabonski, Uli Fahrenberg, Eric Finster, Philippe Gaucher, Rob van Glabeek, Marco Grandis, Tobias Heindel, Maurice Herlihy, Kathryn Hess Bellwald, Mateusz Juda, Philippe Malbos, Nicolas Ninin, Sergio Rajsbaum, Christine Tasson, Krzysztof Worytkiewicz, and Krzysztof Ziemiański for the stimulating discussions without which this book could certainly not have been written, and especially Jérémy Dubut, Sanjeevi Krishnan, and Tim Porter for a careful proofreading of the book and many comments. We would also acknowledge support from AS CNRS TAPESC, ACI project GEOCAL, ANR projects INVAL, CHOCO, and CATHRE, and from the ESF research networking program ACAT.

Paris   Samuel Mimram
September 2015

Contents

Chapter 1
Introduction

Concurrent programs consist of multiple processes running in parallel. Their use has become more and more widespread in order to efficiently exploit recent architectures (processors with many cores, clouds, etc.), but they are notoriously difficult to design and to reason about: one has to ensure that the program will not go wrong, regardless of the way the different processes constituting the program are scheduled. In principle, in order to achieve this task with the help of a computer, we could apply traditional verification techniques for sequential programs on each of the possible executions of the program. But this is not feasible in practice because the number of those executions, or *schedulings*, may grow exponentially with the size of the program. Fortunately, it can be observed that many of the schedulings are equivalent in the sense that one can be obtained from the other by permuting independent instructions: such equivalent executions will always lead to the same result. Hence, if one of those executions can be shown not to lead to an error, neither will any other execution which is equivalent to it.

This suggests that a model for concurrent programs should incorporate not only the possible executions of the program (as in traditional interleaving semantics), but also the commutations between instructions, following the principle of what is now called *true concurrency*. Interestingly, the resulting models are algebraic structures which can be interpreted *geometrically*: roughly as topological spaces in which paths correspond to executions and two executions are equivalent when the corresponding paths are homotopic, i.e., connected by a continuous deformation from one to the other. In order to make this connection precise, it turns out that topological spaces are not exactly the right notion for our purposes. One needs to use a *directed* variant, i.e., to incorporate a notion of irreversible time.

Mutatis mutandis, starting from very practical motivations (the verification of concurrent programs), questions of a more theoretical nature arise. What is a good notion of a directed space, and how do classical techniques from algebraic topology apply to this setting? What is the geometry of concurrent programs? How can a geometrically refined understanding of concurrency be used in order to design new

© Springer International Publishing Switzerland 2016

L. Fajstrup et al., *Directed Algebraic Topology and Concurrency*,
DOI 10.1007/978-3-319-15398-8_1

and more efficient algorithms for studying concurrent programs? The goal of this book is to give a general overview of our current understanding, regarding these questions.

Models for Concurrency

Historically, the first models for concurrent programs were the so-called *interleaving* models, which essentially consist of all sequences of actions that could occur in the execution of a program. For instance, consider a program of the form $A \parallel B$, consisting of two instructions A and B executed in parallel. Its semantics would be the following graph with four vertices and four edges:

$$
\begin{array}{ccc}
y_2 \bullet & \xrightarrow{\quad A \quad} & \bullet z \\
\uparrow & & \uparrow \\
B \Big| & & \Big| B \\
x \bullet & \xrightarrow[\quad A \quad]{} & \bullet y_1
\end{array}
$$

(1.1)

Notice that the two maximal paths are labeled by $A \cdot B$ and $B \cdot A$, i.e., the two interleavings of A and B.

This semantics does not take into account when two sequences of instructions are equivalent, i.e., when the two actions A and B are independent. For instance, with A and B being respectively x:=1 and y:=2, the two actions are independent because any execution of the program will lead to a state where the variables x and y respectively contain 1 and 2. However, this is not the case when A and B are, respectively, x:=2 and x:=2*x. Starting from a state where x contains 0, the execution $A \cdot B$ will end in a state where x contains 4, while execution $B \cdot A$ will end in a state where x contains 2. Even worse, the simultaneous execution of A and B can even end in other states, as sometimes happens in practice. In this case, the order in which the actions are scheduled matters. In order to distinguish between the two cases, we will equip our graph with a relation $\sim$ on paths, indicating when they are equivalent in this sense, in order to obtain what is called an *asynchronous graph*.

In order to avoid two incompatible instructions being executed at the same time, most operating systems provide *mutexes*, particular kinds of resources which can be held by at most one process at a time: given a mutex a, a process can either lock or release the resource by respectively performing the instructions P_a or V_a, and if a process tries to lock a mutex which was already taken, it is then frozen until the mutex is released. From the point of view of the semantics, the usage of those instructions has two effects: first, it forbids some states (those in which more than one process would have locked the mutex), and second it explicitly states that some schedulings are not equivalent. For instance, the semantics of $(A_1 ; A_2 ; A_3) \parallel (B_1 ; B_2 ; B_3)$ and $(P_a ; A ; V_a) \parallel (P_a ; B ; V_a)$ are, respectively,

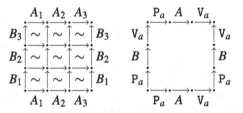

Any two maximal paths in the first graph are equivalent, while the two maximal paths in the second graph are not equivalent. Moreover, the second graph can be obtained from the first one by removing vertices in the middle and adjacent edges (those vertices would correspond to positions where the mutex a is locked twice): we will see that a semantics of programs with mutexes can be generally obtained in this way, by associating an asynchronous graph to a program, and then removing forbidden vertices.

The Geometry of Concurrent Programs

In the asynchronous graph semantics presented above, the executions of the program correspond to paths in the graph. Moreover, the squares where the paths in the boundary are equivalent (i.e., those squares marked with "$\sim$") can be regarded as "filled squares" and the other ones as "empty squares": intuitively, when a square of the form (1.1) is filled, there is enough room to allow for a deformation, or *homotopy*, to exist between paths $A \cdot B$ and $B \cdot A$. In order to make this intuition more formal, it is tempting to investigate another type of geometric model for programs, based on topological spaces instead of graphs, in which an execution corresponds to a path and an equivalence corresponds to a homotopy between paths. For instance, to the program $(\texttt{P}_a \,;\, \texttt{P}_b \,;\, \texttt{V}_b \,;\, \texttt{V}_a) \, \| \, (\texttt{P}_b \,;\, \texttt{P}_a \,;\, \texttt{V}_a \,;\, \texttt{V}_b)$ is associated the topological space on the left below, obtained from $[0, 1] \times [0, 1]$ by removing the darkened region (the points in this region would correspond to the states where either a or b has been locked twice, which is forbidden):

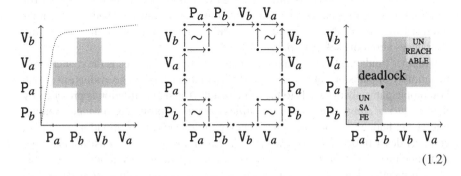

$$(1.2)$$

In this space, the paths starting from the beginning position (the lower left corner) and which are "increasing" (i.e., going right and up) correspond to executions. For instance, the dotted path corresponds to the second process executing $\texttt{P}_b \cdot \texttt{P}_a \cdot \texttt{V}_a$; then the first process executing $\texttt{P}_a$; then the second process executing $\texttt{V}_b$; and finally, the first process executing $P_b . V_b . V_a$. Paths which are not increasing make no sense from a computational point of view: they correspond to executions which go backwards in time at some point. We thus have to consider a variant of the notion of topological space which is *directed* in the sense that the space comes equipped with a time direction, i.e., extra structure specifying which paths can be considered as "increasing" or "directed". One can then study the geometry of these spaces, and in particular the structure of directed paths up to a suitable notion of homotopy, which corresponds to equivalence classes of executions, up to commutation of independent actions, by adapting classical constructions from algebraic topology. The topological semantics can also be precisely related to the asynchronous semantics: notice the similarity of the topological space on the left with the asynchronous graph in the middle! If the situation seems to be quite simple and clear in the above examples, many subtleties occur when more than two processes are involved, i.e., when considering spaces of dimension greater than 2.

Verification of Concurrent Programs

One of the main interests in the connection between semantics of concurrent programs and algebraic topology is that algebraic topology provides one with a new point of view on those programs, thus allowing for the formulation of new algorithms for program verification. For instance, consider the rightmost state space in (1.2). Illustrated is a *deadlock* point. Starting from this point, there exists no non-constant increasing path; in other words, the point corresponds to a state of the program in which no instruction can be executed. This kind of undesirable behavior is specific to concurrent programs, and typically occurs when processes are waiting for each other (e.g., to free a resource or to produce data). The points in the lower left square are called *unsafe*: they correspond to states of the program from which an execution might lead to a deadlock. The points in the upper right square are called *unreachable*: no directed path from the beginning position ends in that square, which indicates the existence of states which can never occur during an execution. While this is not an error per se, their presence is often the sign of a poor design in the program (or worse). Based on the geometric characterization of such states (and others of similar interest), we will be able to formulate algorithms to compute them, thus providing guarantees about the safety of programs.

Another fundamental application of the geometric techniques is the reduction of the number of paths or states to explore, based on the idea that the evaluation of two homotopic paths always leads to the same result. A first construction is provided by the *category of components*, which identifies portions of programs in which "nothing interesting happens" from the concurrency point of view, thus providing us with a compact description of the geometry of the program. A second construction is the computation of the *path space* (the space of directed paths up to homotopy): once this space is computed, it suffices to apply traditional (sequential) verification techniques on one representative of each homotopy class of paths, in order to cover all possible schedulings of the program.

Plan of the Book

We begin by introducing a toy programming language, provide its interleaving semantics, and describe the properties of programs we are interested in (Chap. 2). We then add resources, such as mutexes in the language, and provide truly concurrent semantics for programs; such semantics at first involve asynchronous graphs, and later generalize to precubical sets (Chap. 3). The notion of a directed topological space is introduced and used to provide new semantics, and we discuss a suitable notion of homotopy between paths in the resulting models (Chap. 4). Algorithms based on the geometric semantics are then described for computing: cubical regions, deadlocks, and factorizing processes (Chap. 5). The two next chapters discuss more advanced topics: categories of components (Chap. 6) and paths spaces (Chap. 7). Finally, we conclude by hinting at topics not covered in the book and future developments (Chap. 8).

Reader's Guide

This book is intended both for mathematicians and computer scientists. Mathematicians not accustomed to concurrent languages and their semantics should spend some time reading Chaps. 2 and 3, whereas computer scientists might want to skip most of the standard material up to Sect. 3.4. Similarly, mathematicians (in particular algebraic topologists) can skip the beginning of Sect. 4.2 recalling classical concepts in algebraic topology, which are later on adapted in the directed setting. Chapters 2–4 constitute the syntactical and theoretical core of the book. Subsequent chapters can be mostly read independently:

- Chapter 5 presents algorithms (for representing regions, computing deadlocks, and factorizing processes);
- Chapter 6 introduces a notion of "connected components" in directed topology, which can be used to obtain a compact representation of the category of directed paths up to homotopy;
- and Chap. 7 constructs combinatorial models for the space of directed paths with fixed endpoints up to homotopy, which lead to efficient computations.

These last two chapters, definitely the most mathematical, can be skipped by computer scientists on a first reading. This book is still only an introduction to the subject, with only hints of practical applications, and no detailed proofs. Readers interested in details and further developments will find references in historical sections at the end of the chapters and in Chap. 8, which briefly mentions related approaches that could not be developed here.

Notations

Some mathematical notation will be consistently used throughout the book. We write: $\mathfrak{P}(X)$ for the powerset of a set X; $[1 : n]$ for the set $\{1, \ldots, n\}$; $]x, y[$ (resp. $[x, y]$) for open (resp. closed) intervals (following French convention); $]x, y]$ for intervals which can be either open or closed on both sides; and $f : x \twoheadrightarrow y$ for a path with x as source and y as target.

Chapter 2
A Toy Language for Concurrency

Since the aim of this book is to introduce models and verification techniques for programming languages, our first task is to introduce the programming language through which we demonstrate the main ideas of this book. There are many possible choices for such a language, from theoretical ones (e.g., CCS, the π-calculus [124]) which abstract away implementation details, to real-world standards and languages (e.g., POSIX, Java) with large sets of tools to handle concurrency. We choose to invent an intermediate language which is relatively concise while being somewhat realistic. We begin by introducing the language (Sect. 2.1). We then describe its operational semantics, which formalizes the way programs are to be executed (Sect. 2.2). Finally, we describe the correctness properties that we will be interested in (Sect. 2.3). In this chapter, the execution model of programs running in parallel is formalized in the simplest possible way, as an interleaving of the actions of the programs, and will be refined in the next chapter by truly concurrent models.

2.1 A Toy Language

Throughout the book, we consider a concurrent, shared-memory, imperative, toy language for illustrative purposes. A program in this language consists of a sequence of instructions, as in the following example:

$$x := 3; x := x + 1; y := 2 * x$$

The above sequence first assigns the value 3 to x, then increments x, and finally assigns twice the value of x to y; here x and y are variables representing memory cells which are supposed to contain integers. Such sequences of instructions can be combined, using control flow constructs (e.g., conditional branching, conditional loops), as in the following example:

© Springer International Publishing Switzerland 2016

L. Fajstrup et al., *Directed Algebraic Topology and Concurrency*,

DOI 10.1007/978-3-319-15398-8_2

```
x := 5;
while x != 1 do (
    if x mod 2 != 0 then
        (x := 3*x;  x := x+1)
    else
        x := x/2
);
print "Reached 1!"
```
(2.1)

This program computes the elements of the *Syracuse sequence*—the smallest sequence starting with $x_0 = 5$, ending with 1, and inductively defined for $n > 0$ by $x_{n+1} = x_n/2$ for x_n even and $x_{n+1} = 3x_n + 1$ for x_n odd—before printing a message. Here, $x \ != \ 1$ denotes the condition $x \neq 1$. Such a program can be represented graphically by its *control flow graph*:

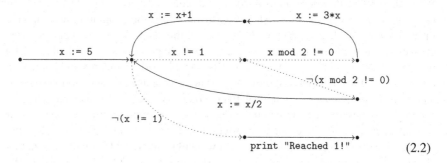

(2.2)

The vertices correspond to the positions in the program (a position is roughly a line number in the code), the solid arrows correspond to instructions, and the dotted arrows correspond to branches which might be taken depending on a condition. An *execution* of the program can be described as a path in this graph starting from the leftmost vertex.

The language we will use is a variant of the IMP language, often used as a setting for studying semantics [166], extended with a parallel composition operator. We choose to illustrate our methods on an imperative language because those are the most widespread, but they could be adapted to other flavors of programming languages (functional, object-oriented, etc.).

Definition 2.1 We suppose a fixed countable set *Var* of variables, and below x denotes a variable and n an integer. The language **PIMP** (*parallel* IMP) comprises three kinds of syntactic expressions, defined by their grammar:

- the set $\mathscr{A}$ of *arithmetic expressions*:

$$a \ ::= \ x \ | \ n \ | \ a + a \ | \ a * a$$

- the set $\mathscr{B}$ of *Boolean expressions*, or *conditions*:

$$b \ ::= \ \text{true} \ | \ \text{false} \ | \ a < a \ | \ b \ \text{and} \ b \ | \ \neg b$$

- the set $\mathscr{C}$ of *commands*, or *programs*:

$$c \ ::= \ x \ := \ a \ | \ \text{skip} \ | \ c; \ c \ | \ \text{if} \ b \ \text{then} \ c \ \text{else} \ c \ |$$
$$\text{while} \ b \ \text{do} \ c \ | \ c \ \| \ c$$

A command of the form "$x \ := \ a$" is called an *action* and we write $\mathscr{C}_{\text{act}}$ for the set of actions. A program not containing the instruction "$\|$" is called *sequential*.

A program is meant to be executed in an environment (the state of the memory) consisting of the values for the variables, which are integers only (for simplicity, we do not consider variables containing Booleans or other data types). Arithmetic and boolean expressions evaluate to integers and Booleans, respectively, in the usual way (e.g. , $*$ refers to integer multiplication, $\neg$ refers to logical negation). The commands have an effect on memory or on the control flow of the program. Their respective meanings are given in table below.

`x := a`	assign the result of the evaluation of the arithmetic expression a to the variable x
`skip`	do nothing
`c₁; c₂`	sequentially execute c_1 and then c_2
`if b then c₁ else c₂`	branch conditionally, i.e., evaluate the Boolean expression b and execute c_1 (resp. c_2) if the result is true (resp. false)
`while b do c`	execute the command c as long as the Boolean expression b evaluates to true
`c₁ ‖ c₂`	execute c_1 in parallel with c_2

We refer the reader to standard textbooks [166] for details about the first five operations, standard constructs in sequential programs. We will later discuss extensively the last, and novel, operation "$\|$" of *parallel composition*. The language is deliberately small in order to ease the definitions, but other operators can easily be added, and from time to time we incorporate such operators in our examples without comment, e.g. , the arithmetic operator mod and the Boolean operator $!=$ in Example (2.1).

Convention 2.2 The sequence operator ";" binds more tightly than the parallel operator "$\|$": the program $A; B\|C$ is implicitly parenthesized as $(A; B)\|C$. Moreover, multiple sequences or multiple parallels are parenthesized on the right: $A; B; C$ means $A; (B; C)$, $A\|B\|C$ means $A\|(B\|C)$, etc. This last convention is not really important though, because these operators are essentially associative; see Proposition 2.25.

We will focus on the study of *concurrent programs*, in which many subprograms (also called *threads* or *processes*) run in parallel. Programs of such form are often used in order to efficiently exploit the computing resources at our disposal and or be more reactive to external events. A prototypical example is image processing on a multi-core machine, where each thread processes a part of the image for speed. Another prototypical example is the control of a power plant, where one thread controls physical hardware, another thread handles interactions between hardware and operators, and a failure of the latter does not imperil the former. Most common operating systems provide facilities for dynamic thread creation and associated operations, such as those formalized in the POSIX standard [79]. Once again, we abstract away implementations details. Instead we include the operation || in the very definition of a program; $p_1 || p_2$ means that the programs p_1 and p_2 are run in parallel, typically as two different threads. For instance, a simple image processing program might look like

$$p_i; \ (p_l || p_r) ; \ p_d$$

where p_i takes care of the initialization of the program, p_l and p_r process, respectively, the left and right parts of the image, and p_d displays the resulting image.

As a first approximation, the effect of $p || q$ is the same as some interleaving of actions of p and q: "the result of any execution is the same as if the operations of all the processors were executed in some sequential order, and the operations of each individual processor appear in this sequence in the order specified by its program", as phrased by Lamport [110]. We elaborate on the validity of this assumption, called *sequential consistency*, in Remark 2.17.

Remark 2.3 We limit ourselves in this book to the semantics of finitely many threads, whereas in most languages there are ways to define threads recursively, potentially creating an unbounded number of threads. In many ways, going to that level of generality would obscure the main purpose of the book without covering many more applications in practice: many programs are essentially structured as the image processing example above, creating all the threads after an initialization phase. To give a simple idea of the difference made by adding recursive thread creation to a parallel language, let us just mention that the state reachability problem for (our) multithreaded programs is PSPACE complete when there are finitely many finite-state threads [101] and undecidable with recursive threads [144].

2.2 Semantics of Programs

In this section, we formally introduce the notion of a transition graph (or a control flow graph) associated to a program [1]. This classical construction allows one to abstract away from the syntax of the programming language, to easily define a notion of execution trace for a program, and to provide an operational semantics.

2.2.1 Graphs

We begin by recalling some well-known constructions on graphs.

Definition 2.4 A *graph* $G = (V, \partial^-, \partial^+, E)$ consists of a set V of *vertices* (or *states*), a set E of *edges* (or *transitions*), and two functions $\partial^-, \partial^+ : E \to V$, respectively, associating to an edge its *source* and its *target*.

When graphs are indexed (such as G_1), we often index the associated sets of vertices (V_1), edges (E_1), and source and target functions (∂_1^- and ∂_1^+) correspondingly.

Definition 2.5 Given a set $\mathscr{L}$ of *labels*, a *labeled graph* (G, ℓ) consists of a graph G, as in Definition 2.4, together with a function $\ell : E \to \mathscr{L}$. Given an edge $e \in E$, the element $\ell(e)$ is called the *label* of the edge e.

We sometimes write $x \xrightarrow{A} y$ for an edge e such that $\partial^-(e) = x$, $\partial^+(e) = y$, and $\ell(e) = A$. This notation is not ambiguous because we usually consider graphs in which the edges between a given pair of vertices have different labels. A *path* $t = e_1 . e_2 \ldots e_n$ is a finite nonempty sequence of edges $e_i \in E$ such that $\partial^+(e_i) = \partial^-(e_{i+1})$ for $1 \leq i < n$, the integer n being the *length* of the path, or a vertex x denoting the empty path on this vertex, often written as ε_x. We write $\partial^-(t) = \partial^-(e_1)$ (resp. $\partial^+(t) = \partial^+(e_n)$) for the source (resp. target) of the path. Two paths with the same source (resp. target) are called *coinitial* (resp. *cofinal*). Given two paths t and u such that $\partial^+(t) = \partial^-(u)$, we write $t . u$ for their *concatenation*. Given two vertices $x, y \in E$, we say that y is *reachable* from x when there exists a path t with $\partial^-(t) = x$ and $\partial^+(t) = y$, denoted $t : x \twoheadrightarrow y$. When the graph is labeled in $\mathscr{L}$, the sequence of labels of edges in a path t forms a word in $\mathscr{L}^*$ that we denote as $\ell(t)$.

A *morphism* $f : G_1 \to G_2$ between two graphs consists of a pair of functions $f^V : V_1 \to V_2$ and $f^E : E_1 \to E_2$ such that the function on edges is compatible with source and target:

$$f^V \circ \partial_1^- = \partial_2^- \circ f^E \quad \text{and} \quad f^V \circ \partial_1^+ = \partial_2^+ \circ f^E$$

When the two graphs are labeled with the same set of labels, the morphism is moreover required to preserve the labeling of edges: $\ell_1 = \ell_2 \circ f^E$. The two functions f^V and f^E are often abusively denoted by the same symbol f, the context making clear which one we are referring to. Two (labeled) graphs G_1 and G_2 are isomorphic when there exists morphisms $f : G_1 \to G_2$ and $g : G_2 \to G_1$ such that $g \circ f = \mathrm{id}$ and $f \circ g = \mathrm{id}$.

In the following, we will make frequent use of the following operations in order to combine graphs:

Definition 2.6 Suppose given two labeled graphs $G_1 = (V_1, \partial_1^-, \partial_1^+, E_1, \ell_1)$ and $G_2 = (V_2, \partial_2^-, \partial_2^+, E_2, \ell_2)$. We define the following constructions on G_1 and G_2.

- Their **disjoint union** $G_1 \sqcup G_2$ is the graph

$$G_1 \sqcup G_2 \;=\; (V_1 \sqcup V_2, \partial^-, \partial^+, E_1 \sqcup E_2, \ell)$$

where $\partial^-(e) = \partial_1^-(e)$ if $e \in E_1$ and $\partial^-(e) = \partial_2^-(e)$ if $e \in E_2$, and similarly for ∂^+ and ℓ.

- Their **tensor product** $G_1 \otimes G_2$ is the graph

$$G_1 \otimes G_2 \;=\; (V_1 \times V_2, \partial^-, \partial^+, (E_1 \times V_2) \sqcup (V_1 \times E_2), \ell)$$

with $\partial^-(e, x) = (\partial_1^-(e), x)$ and $\ell(e, x) = \ell(e)$ for an edge $(e, x) \in E_1 \times V_2$, $\partial^-(x, e) = (x, \partial_2^-(e))$ and $\ell(x, e) = \ell(e)$ for an edge $(x, e) \in V_1 \times E_2$, and similarly for ∂^+.

- Given two vertices $x, y \in V_1$ the **quotient** graph $G_1[x = y]$ is the graph obtained by identifying the vertices x and y in G_1, i.e., formally

$$G_1[x = y] \;=\; (V_1/\approx, \partial^-, \partial^+, E_1, \ell_1)$$

where $V_1/\approx$ is the quotient of the set V_1 by the equivalence relation $\approx$ on V_1 such that $x' \approx x''$ whenever $x' = x''$, or $x' = x$ and $x'' = y$, or $x' = y$ and $x'' = x$; and ∂^- and ∂^+ are the maps induced from ∂_1^- and ∂_1^+ by the quotient.

- Given a subset $V \subseteq V_1$ of vertices of G_1, the **restriction** of G_1 to V is the graph $G_1|_V = (V, E)$ where $E = \{e \in E_1 \mid \partial^-(e) \in V \text{ and } \partial^+(e) \in V\}$.

Example 2.7 From the two graphs

$$G = x \xrightarrow{\;A\;} y \qquad \text{and} \qquad H = z_0 \xrightarrow{\;B_1\;} \cdots$$

we can compute the following graphs:

Notice that the tensor product can be thought of as multiple copies of edges coming from either of the two graphs.

From these operations one can derive most usual operations on graphs. For instance, given a graph G, the graph obtained from G by adding an edge between two vertices x and y is the graph $(G \sqcup I)[x = x'][y = y']$ where I is the graph with two vertices x' and y' and one arrow $x' \to y'$.

Remark 2.8 The tensor product is sometimes referred to as the "cartesian product" of the two graphs. However, this terminology is incorrect since it is *not* a cartesian product in the usual category of graphs (and to add to the confusion, the proper cartesian product is usually called the tensor product). In fact, most of the classical interleaving semantics of such concurrent systems was originally done using yet another variant of product, called *synchronized product* [135]. The definition of the tensor product given above should become more natural when seen as a particular (one-dimensional) case of the tensor product of precubical sets, as defined in Sect. 3.4.

2.2.2 The Transition Graph

The operations introduced in the previous section easily allow us to formalize the notion of transition graph (or control flow graph) associated to a program as follows:

Definition 2.9 The **transition graph** $G_p = (G_p, \ell_p, s_p, t_p)$ associated to a program p is a graph G_p labeled in the set $\mathscr{L} = \mathscr{C}_{\text{act}} \sqcup \mathscr{B}$ together with two distinguished vertices $s_p, t_p \in E$ called the *beginning* and *end*. This graph is defined inductively as follows:

- G_A with $A \in \mathscr{C}_{\text{act}}$ is the graph with two vertices and one edge labeled by A:

$$s_A \bullet \xrightarrow{\quad A \quad} \bullet t_A$$

- G_{skip} is the graph with one vertex (being both the beginning and the end) and no edge:

$$s_{\text{skip}} \bullet t_{\text{skip}}$$

- $G_{p;q}$ is the graph obtained from the disjoint union of G_p and G_q by identifying t_p with s_q, such that $s_{p;q} = s_p$ and $t_{p;q} = t_q$:

$$s_{p;q} = s_p \bullet \quad G_p \quad t_p \bullet s_q \quad G_q \quad \bullet t_q = t_{p;q}$$

- G_p, with $p = \text{if } b \text{ then } p_1 \text{ else } p_2$, is the graph obtained from the disjoint union of G_{p_1} and G_{p_2} by identifying t_{p_1} and t_{p_2}, the resulting vertex being t_p, adding a new vertex s_p and two transitions $s_p \xrightarrow{b} s_{p_1}$ and $s_p \xrightarrow{\neg b} s_{p_2}$:

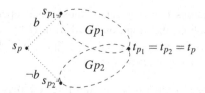

- G_p, with p=while b do q, is obtained from G_q by adding a vertex t_p, adding an edge $t_q \xrightarrow{\neg b} t_p$, and adding an edge $t_q \xrightarrow{b} s_q$, with $s_p = t_q$:

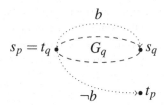

(notice that source of the graph G_q is on the right and the target on the left in the above figure)
- $G_{p\|q}$ is the graph $G_p \otimes G_q$ with $s_{p\|q} = (s_p, s_q)$ and $t_{p\|q} = (t_p, t_q)$.

A vertex of G_p is called a **position** of the program p.

Notice again that all the graphs above can be obtained using the constructions of Definition 2.6, for instance $G_{p;q} = (G_p \sqcup G_q)[t_p = s_q]$. As shown above and in the introductory example (2.1), the edges labeled by conditions in transition graphs are drawn with dotted arrows to distinguish them from those labeled by actions. This is only a drawing convention; there is no difference between the two types of edges except the sets in which they are labeled.

Example 2.10 The Syracuse program (2.1) gives rise to the transition graph depicted in (2.2).

Example 2.11 The transition graph of the program $p = (A; B; C) \| (D; E)$, where A, B, C, D, E are arbitrary actions, is

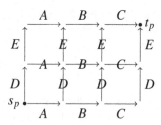

By construction of the transition graph, we have

Lemma 2.12 *For every program p and vertex x of its transition graph there are, by construction, both a path from s_p to x and a path from x to t_p.*

Paths starting at the beginning vertex will be of particular importance since they encompass all the sequences of actions that the program can give rise to.

Definition 2.13 A **potential execution trace** of a program p is a path starting from s_p in G_p. We write $T_{pot}(p)$ for the set of such paths.

Of course, not every such path corresponds to an actual execution of the program, which is why they are called "potential." Determining the ones which can actually occur during an execution depends on the chosen semantics of the programming language, as formalized in Definition 2.19. Notice that the set $T_{pot}(p)$ is closed under prefix, its maximal elements are thus enough to describe it when traces are of bounded length.

Example 2.14 Consider a program p of the form A; (if b then B else C). Its maximal potential execution traces are labeled by $A.b.B$ and $A.\neg b.C$. The conditions occurring in those traces should be thought of as assumptions on the memory state under which those traces make sense, which is why we call them *potential*. For instance, the trace $A.b.B$ should be read as: do A, now suppose that the condition b is true, do B.

Example 2.15 Given an action A, the potential execution traces of the program while b do A are labeled by words which are prefixes of words in $(b.A)^*.\neg b$. In particular, when the condition b is true, all those labeled in $(b.A)^*$ are valid: a program can thus admit an infinite number of traces.

Example 2.16 Consider again the program $p = (A; B; C)||(D; E)$ introduced in Example 2.11. Its maximal potential execution traces are labeled by elements of the set $\{A.B.C\} \sqcup \{D.E\}$ of shuffles of the words $A.B.C$ and $D.E$, i.e.,

$$\{ A.B.C.D.E, \ A.B.D.C.E, \ A.D.B.C.E, \ D.A.B.C.E, \ A.B.D.E.C,$$
$$A.D.B.E.C, \ D.A.B.E.C, \ A.D.E.B.C, \ D.A.E.B.C, \ D.E.A.B.C \}$$

Given two languages $P, Q \subseteq \mathscr{L}^*$, we recall that their *shuffle* $P \sqcup Q$ is defined inductively by

$$P \sqcup Q \ = \ \bigcup_{a \in \mathscr{L}} \{a\} . ((P/a \sqcup Q) \cup (P \sqcup Q/a))$$

with $P/a = \{u \in \mathscr{L}^* \mid a.u \in P\}$ and similarly for Q/a. Said otherwise, the words $u = a_1 \ldots a_n$ in $P \sqcup Q$ are those for which there exists a subset $I \subseteq \{1, \ldots, n\}$ such that the subword of u consisting of letters with indices in I (resp. in $\{1, \ldots, n\} \setminus I$) is in P (resp. in Q). This example illustrates why we chose to interpret the parallel composition by the tensor product of graphs in Definition 2.9: in order to execute $p||q$, one should either execute the first action of p and then the rest of p in parallel with q, or the first action of q and then p in parallel with the rest of q. It is also interesting to notice the large number of execution traces, considering the small size of the program generating them: a word in the above shuffle is of length $3 + 2 = 5$, and is characterized by the positions of the subword of length 2 in this word, so the cardinality of the above set is $\binom{5}{2} = 5!/(2!\,3!) = 10$.

Remark 2.17 The assumption that the only behaviors that can occur in a concurrent program are those which can be obtained as a sequential interleaving of the instructions of the processes executed in parallel is called *sequential consistency* [110]. Real multiprocessors, however, use sophisticated techniques to achieve high performance: the storage of buffers, hierarchies of local cache, speculative execution [154], etc. These implementation details are not observable by single-threaded programs, but in multithreaded programs different threads may see subtly different views of memory. Such machines exhibit *relaxed* (or *weak*) *memory models*. For instance, consider a standard x86 processor. Given two memory locations x and y (initially holding the value 0), we look at the following program with two threads writing 1 to both x and y and then reading from y and x:

$$(\texttt{mov x, 1 ; mov eax, y}) \quad || \quad (\texttt{move y, 1 ; mov ebx, x})$$

The instruction "mov x, y" is essentially the assembly notation for $x := y$, and eax and ebx are special memory locations called registers. Intuitively, the possible outcomes for (eax, ebx) are (1, 1), (1, 0), and (0, 1). However, on standard processors, some executions can also lead to (0, 0) [131]. Throughout the book we still make the assumption that the semantics is sequentially consistent for simplicity (our approach could be refined to encompass a semantics with a relaxed memory model) and because modern compilers help to ensure sequentially consistent semantics at a higher level.

2.2.3   Operational Semantics

We now need to describe, formally, the effect of executing a program, by describing a semantics for the language that we have been specifying. Notice that there can be many different semantics for a given language: most of the following constructions depend on our choice, and would be (slightly) different if we chose a different semantics. A program is meant to be executed in a *state* (also sometimes called an *environment*), which comprises the contents of the memory (as well as other resources to which the program would have access). For instance, the effect of an action like x:=x+1 is to modify the memory cell corresponding to the variable x by incrementing x. Similarly, given a memory state, a condition such as x<1 can be evaluated to either true or false depending on whether the cell x contains a value below 1 or not. In the following, we write $\mathbb{B} = \{\perp, \top\}$ for the set of *Booleans*, where $\perp$ (resp. $\top$) denotes false (resp. true).

Definition 2.18 We write $\Sigma = \mathbb{Z}^{Var}$ for the set of *states*, consisting of functions assigning an integer to each variable. The *initial state* $\sigma_0 \in \Sigma$ is the constant function equal to 0. The **operational semantics** of our programming language consists of three functions:

- $[\![-]\!]_{\mathscr{A}} : \mathscr{A} \to (\Sigma \to \mathbb{Z})$ describing the evaluation of arithmetic expressions,
- $[\![-]\!]_{\mathscr{B}} : \mathscr{B} \to (\Sigma \to \mathbb{B})$ describing the evaluation of Boolean expressions,
- $[\![-]\!]_{\mathscr{C}^*_{\mathrm{act}}} : \mathscr{C}^*_{\mathrm{act}} \to (\Sigma \to \Sigma)$ describing the effect of sequences of actions on the state.

Thus, for instance, $[\![-]\!]_{\mathscr{A}}$ sends an element of $\mathscr{A}$ (an arithmetic expression) to a function from the set Σ of states to the set of integers. Given an arithmetic expression a, we write $[\![a]\!]_{\mathscr{A}} : \Sigma \to \mathbb{Z}$ for its semantic interpretation, and similarly for the other functions. Given a state $\sigma \in \Sigma$, the evaluation of arithmetic expressions is defined by

$$[\![x]\!]_{\mathscr{A}}(\sigma) = \sigma(x) \qquad [\![n]\!]_{\mathscr{A}}(\sigma) = n \qquad [\![a_1{+}a_2]\!]_{\mathscr{A}}(\sigma) = [\![a_1]\!]_{\mathscr{A}}(\sigma) + [\![a_2]\!]_{\mathscr{A}}(\sigma) \qquad \ldots$$

the evaluation of Boolean expressions by

$$[\![\mathtt{true}]\!]_{\mathscr{B}}(\sigma) = \top \qquad [\![\mathtt{false}]\!]_{\mathscr{B}}(\sigma) = \bot$$

$$[\![a_1{<}a_2]\!]_{\mathscr{B}}(\sigma) = \begin{cases} \top & \text{if } [\![a_1]\!]_{\mathscr{A}}(\sigma) < [\![a_2]\!]_{\mathscr{A}}(\sigma) \\ \bot & \text{otherwise} \end{cases} \qquad \ldots$$

and the evaluation of sequences of actions in $\mathscr{C}^*_{\mathrm{act}}$ by

$$[\![u \,.\, v]\!]_{\mathscr{C}^*_{\mathrm{act}}}(\sigma) = [\![v]\!]_{\mathscr{C}^*_{\mathrm{act}}} \circ [\![u]\!]_{\mathscr{C}^*_{\mathrm{act}}}(\sigma) \qquad [\![\varepsilon]\!]_{\mathscr{C}^*_{\mathrm{act}}}(\sigma) = \sigma$$

$$[\![x := a]\!]_{\mathscr{C}^*_{\mathrm{act}}}(\sigma) = \sigma\left[x \mapsto [\![a]\!]_{\mathscr{A}}(\sigma)\right] \qquad \ldots \tag{2.3}$$

where ε denotes the empty word and, given $n \in \mathbb{Z}$, $\sigma[x \mapsto n]$ denotes the function in Σ which associates n to x and $\sigma(y)$ to each $y \neq x$.

In the following, we generally drop the subscripts when the function to which we are referring is clear from context. The operational semantics allows us to characterize those potential execution traces which are valid, in the sense that the assumptions corresponding to the conditions occurring in those traces are satisfied. Given a word u in $(\mathscr{C}_{\mathrm{act}} \sqcup \mathscr{B})^*$, we write $[\![u]\!] : \sigma \to \sigma$ defined as in (2.3), extended with $[\![b]\!](\sigma) = \sigma$, for a condition $b \in \mathscr{B}$, in other words $[\![u]\!] = [\![v]\!]_{\mathscr{C}^*_{\mathrm{act}}}$ where v is the projection of u onto $\mathscr{C}^*_{\mathrm{act}}$. Finally, given a path t in a transition graph G_p (see Definition 2.9), its labeling word $\ell(t)$ is an element of $(\mathscr{C}_{\mathrm{act}} \sqcup \mathscr{B})^*$, and we sometimes write $[\![t]\!]$ instead of $[\![\ell(t)]\!]$. In particular, notice that $[\![t]\!]$ is defined for each potential execution trace $t \in \mathrm{T}_{\mathrm{pot}}(p)$, even if it is not valid.

Definition 2.19 A word $u \in (\mathscr{C}_{\mathrm{act}} \sqcup \mathscr{B})^*$ is *valid* when it is either

- the empty word ε,
- of the form $u \,.\, A$ with u valid and $A \in \mathscr{C}_{\mathrm{act}}$,
- or of the form $u \,.\, b$ with u valid, $b \in \mathscr{B}$ and $[\![b]\!] \circ [\![u]\!](\sigma_0) = \top$.

An **execution trace** is a valid potential execution trace in $T_{pot}(p)$, i.e., a path $t \in T_{pot}(p)$ such that the word $\ell(t)$ is valid. We write $T(p)$ for the set of execution traces of a program p.

Example 2.20 Consider the following program:

$$(x := 0 \ \| \ x := 1); \quad \text{if} \ \ x == 0 \ \ \text{then} \ \ c_0 \ \ \text{else} \ \ c_1$$

where the $==$ operator compares two integer values for equality. Its four maximal potential execution traces are

$$\begin{array}{ll} \texttt{x:=0.x:=1.x==0}.c_0 & \texttt{x:=0.x:=1.}\neg\texttt{x==0}.c_1 \\ \texttt{x:=1.x:=0.x==0}.c_0 & \texttt{x:=1.x:=0.}\neg\texttt{x==0}.c_1 \end{array}$$

and only traces up to the right and below to the left are valid. For instance, the one up to the left is not valid because when the condition x==0 is evaluated, the variable x contains 1, and the condition is not true. Notice that this example shows that our programs are nondeterministic because of parallelism: here, either c_0 or c_1 can be executed. It can be shown, however, that programs without parallelism are deterministic.

The operational semantics, as we formulated them, can be related to a more standard *small-step operational semantics* for the programming language. We describe them here briefly and refer to [166] for details about such definitions.

Definition 2.21 We define a **reduction** relation $\rightarrow$ on pairs $\langle \sigma, c \rangle$ consisting of a state $\sigma \in \Sigma$ and a command c, which formally describes how a command evaluates in a given environment. The rules defining this relation are as follows, where each "fraction" below should be interpreted so that the denominator also holds whenever the numerator holds.

$$\frac{\langle \sigma, c_1 \rangle \rightarrow \langle \sigma', c_1' \rangle}{\langle \sigma, c_1 \,;\, c_2 \rangle \rightarrow \langle \sigma', c_1' \,;\, c_2 \rangle} \qquad \frac{}{\langle \sigma, \texttt{skip}\,;c \rangle \rightarrow \langle \sigma, c \rangle} \qquad \frac{}{\langle \sigma, x := a \rangle \rightarrow \langle \sigma[x \mapsto \llbracket a \rrbracket(\sigma)], \texttt{skip} \rangle}$$

$$\frac{\llbracket b \rrbracket(\sigma) = \mathsf{T}}{\langle \sigma, \texttt{if } b \texttt{ then } c_1 \texttt{ else } c_2 \rangle \rightarrow \langle \sigma, c_1 \rangle} \qquad \frac{\llbracket b \rrbracket(\sigma) = \bot}{\langle \sigma, \texttt{if } b \texttt{ then } c_1 \texttt{ else } c_2 \rangle \rightarrow \langle \sigma, c_2 \rangle}$$

$$\frac{\llbracket b \rrbracket(\sigma) = \mathsf{T}}{\langle \sigma, \texttt{while } b \texttt{ do } c \rangle \rightarrow \langle \sigma, c ; \texttt{while } b \texttt{ do } c \rangle} \qquad \frac{\llbracket b \rrbracket(\sigma) = \bot}{\langle \sigma, \texttt{while } b \texttt{ do } c \rangle \rightarrow \langle \sigma, \texttt{skip} \rangle}$$

$$\frac{\langle \sigma, c_1 \rangle \rightarrow \langle \sigma', c_1' \rangle}{\langle \sigma, c_1 \| c_2 \rangle \rightarrow \langle \sigma', c_1' \| c_2 \rangle} \qquad \frac{\langle \sigma, c_2 \rangle \rightarrow \langle \sigma', c_2' \rangle}{\langle \sigma, c_1 \| c_2 \rangle \rightarrow \langle \sigma', c_1 \| c_2' \rangle}$$

$$\frac{}{\langle \sigma, \texttt{skip} \| c \rangle \rightarrow \langle \sigma, c \rangle} \qquad \frac{}{\langle \sigma, c \| \texttt{skip} \rangle \rightarrow \langle \sigma, c \rangle}$$

The relation between the semantics given in Definition 2.18 and the one of Definition 2.21 can be formalized as follows. We write $\rightarrow^*$ for the reflexive and transitive closure of $\rightarrow$.

Proposition 2.22 *Given states $\sigma, \sigma' \in \Sigma$ and a command c, there is an execution trace $t \in T(c)$ such that $\llbracket t \rrbracket(\sigma) = \sigma'$ if and only if there exists a command c' such that $\langle \sigma, c \rangle \rightarrow \langle \sigma', c' \rangle$.*

Remark 2.23 In order to show the previous proposition, it is important to remark that in Definition 2.21 the evaluation of arithmetic and Boolean conditions is "atomic", i.e., performed at once. Because of this, the two arithmetic expressions x+x and 2*x are equivalent, in the sense that replacing one with another in a program does not change the results of the program. This would not be true anymore if we had chosen a small-step semantics for the evaluation of expressions as well, i.e., if we had added the rules

$$\frac{\langle \sigma, a_1 \rangle \rightarrow \langle \sigma', a_1' \rangle}{\langle \sigma, a_1 + a_2 \rangle \rightarrow \langle \sigma', a_1' + a_2 \rangle} \qquad \frac{\langle \sigma, a_2 \rangle \rightarrow \langle \sigma', a_2' \rangle}{\langle \sigma, a_1 + a_2 \rangle \rightarrow \langle \sigma', a_1 + a_2' \rangle} \qquad \frac{}{\langle \sigma, n_1 + n_2 \rangle \rightarrow \langle \sigma, n \rangle}$$

with $n_1, n_2 \in \mathbb{Z}$ and $n = n_1 + n_2$, replaced the rule for assignation by the rules

$$\frac{\langle \sigma, a \rangle \rightarrow \langle \sigma', a' \rangle}{\langle \sigma, x := a \rangle \rightarrow \langle \sigma', x := a' \rangle} \qquad \frac{}{\langle \sigma, x := n \rangle \rightarrow \langle \sigma [x \mapsto n], \mathtt{skip} \rangle}$$

with $n \in \mathbb{Z}$, and similarly had replaced the other rules. Under these modified rules, consider the program c defined by y:=(x+x)+1 || x:=x+1. We have $\langle \sigma_0, c \rangle \rightarrow^* \langle \sigma, \mathtt{skip} \rangle$ where σ is a state such that $\sigma(y) = 2$, because the incrementation of x can be interleaved between the two evaluations of the variable x occurring in the expression defining y. This is not possible anymore if we replace x+x with 2*x. However, this kind of behavior could be simulated by introducing variables for intermediate results during the evaluation of expressions.

Finally, we would like to briefly mention the concept of contextual equivalence of programs, applied to concurrent programs.

Definition 2.24 Two commands c_1 and c_2 are *contextually equivalent*, written $c_1 \approx c_2$, when for every state $\sigma \in \Sigma$ we have $[\![c_1]\!](\sigma) = [\![c_2]\!](\sigma)$.

It is well known that, up to contextual equivalence, sequential composition is associative and admits $\mathtt{skip}$ as its neutral element. Similar such identities hold for parallel composition:

Proposition 2.25 *For all commands c, c_1, c_2, and c_3, the following equivalences hold:*

$$(c_1||c_2)||c_3 \approx c_1||(c_2||c_3) \qquad \mathtt{skip}||c \approx c \approx c||\mathtt{skip} \qquad c_1||c_2 \approx c_2||c_1$$

Remark 2.26 In our programming language, the classical equivalence

while b do c $\quad \approx \quad$ if b then $(c;$ while b do $c)$ else skip

can be shown, formalizing the intuition that a while loop can be seen as an infinite sequence of nested conditional branchings. If one is interested in verifying loops of a program only up to a certain depth (i.e., imposing a bound on the number of times a loop can be taken), one can *unroll* the program by replacing every `while` construct by a finite series of conditional branchings as explained above. With this unrolling, we can only verify weaker properties on programs (since we only ensure that the properties are verified up to the depth of the loops). Still, it is sometimes useful since loops are quite difficult to handle in verification.

2.3   Verifying Programs

2.3.1   Correctness Properties

In order to check that a program is "correct," we have to specify what "correctness" means, i.e., what properties we are interested in. We can identify three major families of commonly encountered verification properties:

1. *Functional properties*. These describe how the result of the program complies with a mathematical specification. For instance, an implementation of the factorial function actually computes the factorial, i.e., given an integer $n \in \mathbb{N}$ as input, it returns the integer $n!$. These usually describe invariants or safety properties that will hold true for all executions of the program, expressed in proof-theoretic form [109] or using temporal logic, and generally verified using proof assistants, model checking [27], or abstract interpretation [127].
2. *Reachability properties*. These properties consist in ensuring that some position of a program, typically corresponding to an error, will not be reached. They can also be used to ensure that operations are used within their domain of definition. For instance, division is only defined if the denominator is non-null. In a program, an expression such as `y = 1/x` could be treated as

$$\text{if } x == 0 \text{ then } error \text{ else } y := 1/x \qquad (2.4)$$

   and then a reachability analysis could be used in order to ensure that the error is unreachable (i.e., never executed), which means that x is always non-null at this point of the program. One is also sometimes interested in knowing which positions of a program can be reached, such as in *code coverage* analysis, to ensure that every piece of code can be executed in some situation. Reachability is a particular case of a liveness property [2], and of more general temporal properties as below.
3. *Temporal properties*. They specify the shape of expected execution traces. For instance, in a computer graphics software, whenever the function to display the image is called, the function to compute the image should have been called before.

For simplicity, we are going to focus on reachability properties in the rest of the book: we want to ensure that a given set $\mathscr{X}$ of *forbidden positions* is never reached during the execution of a program. For instance, suppose that there is a special instruction `error`, as in program p shown in (2.4), and we want to ensure that this instruction is never executed, i.e., the position in $\mathcal{X} = \{x_{\text{err}}\}$ in the transition graph corresponding to p is not reachable:

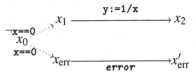

Definition 2.27 A program p is **correct** w.r.t. a set $\mathcal{X}$ of positions if there is no execution trace with a state in $\mathcal{X}$ as target.

In order to verify that a program is correct starting from X_0, we could thus verify all its traces, by exploring all the paths in G_p starting from the initial vertex s_p, following some exploration strategy (e.g. , depth-first, breath-first, etc.). During the exploration, the state $\sigma \in \Sigma$ reached by the path can be iteratively computed following Definition 2.18, and it can be thus checked whether the path is valid (see Definition 2.19), i.e., if it is an execution trace: if the path is valid, the algorithm should check that it does not reach a forbidden position, otherwise the exploration of paths extending the current path can be skipped (since such extended paths will not be valid either).

It can be noticed that the algorithm we have just described, essentially consists of executing the program, with all its possible schedulings, and observing whether an error occurs at some point. This state space exploration strategy is quite naive. For instance, because of `while` loops, the number of traces is not necessarily finite, see for instance, Example 2.15: this problem is not specific to concurrent programs, and traditional methods can be used in order to overcome this shortcoming (one can be interested in correctness properties on execution traces of bounded length and unroll loops, or use widening operators associated to abstract interpretation domains [29], etc.).

In the rest of the book, we will put aside the problems encountered in the verification of generic properties, such as the ones mentioned above, and will focus on those specific to concurrent programs. In this context, one of the main difficulties to overcome is the following one. In order to check that a program is correct, one has to check that all the execution traces coming from possible schedulings of the threads are correct: even without loops, a concurrent program can generate a number of traces which is exponential in the size of the program—namely, by generalizing observations made in Example 2.16! it is easy to see that a program p of the form $p = A||A||\ldots||A$ with n copies of some action A generates $n!$ maximal execution traces. This major problem is often referred to as the *state space explosion problem* [26].

2.3.2 Reachability in Concurrent Programs

The notion of correctness of programs depends on the set $\mathscr{X}$ of positions which we do not want to be reached by any execution. Apart from ensuring that the domain of definition of used operators is respected, such as in the division example (2.4), or more generally that invariants are preserved during computation, sets of positions satisfying the following properties are interesting for concurrent programs: such positions are witnessing potential errors in the code, and the properties are generic in the sense that they are not tied to some particular instructions.

Definition 2.28 A vertex x in the transition graph G_p of a program p is called

- **unreachable** when there is no execution trace with x as target,
- a **deadlock** when x is different from the terminal position t_p and there is an execution trace with x as target which is not a proper prefix of another execution trace,
- **unsafe** when there is an execution trace with x as target which is the prefix of an execution trace with a deadlock as target,
- **doomed** when there is an execution trace with x as target which is not a proper prefix of an execution trace reaching the terminal position t_p.

An unreachable position is a position that can never be reached during the execution of a program: these positions are witnessing the presence of *dead code*, code that will never be executed. In a critical system, every single piece of code is usually written for some purpose, and the fact that some part of the code is formally useless is generally a good indicator of some misconception on the part of the programmer regarding the possible executions of the program. Deadlocks are much more problematic per se: they indicate positions in which the program is blocked and cannot do anything, i.e., the program is "frozen." This situation typically occurs in concurrent programs when two threads (or more) are mutually waiting for each other to free a resource, whence comes the term *deadlock* or *deadly embrace* [31]. Finally, an unsafe position is one from which the program can reach a deadlock position, and a doomed position is one from which the program will eventually reach a deadlock or loop forever. A program p is **safe** when its beginning position s_p is not unsafe: in such a program, no execution will lead to a deadlock.

Remark 2.29 The deadlock and unsafe situations are really specific to concurrent programs (such a program exhibiting a deadlock is provided in Example 3.32): it can be shown that a sequential program does not have deadlock positions.

Remark 2.30 We could have introduced a variant of the notion of deadlock (and similarly for unsafe and doomed positions) by requiring that all the execution traces reaching x cannot be extended. This would have not made any difference in what follows, because we will restrict to programs (called "coherent programs") whose structure is such that a property of a position does not depend on the path reaching it: for those programs either all the paths reaching x cannot be extended, or none.

Remark 2.31 Following the tradition in classical automata theory, we will only consider finite executions. However, even in this case, there are some subtle differences related to the presence of infinite executions (that will be silently ignored in the rest of this book): for instance, a doomed position could have been defined as a reachable position x such that every maximal path with x as source has a deadlock as target. Notice that this definition is not equivalent to the one above for programs with loops.

The search for such positions in concurrent programs will be extensively discussed and illustrated in the next section (in particular, Examples 3.22 and 3.23 provide programs illustrating positions with these properties).

Chapter 3
Truly Concurrent Models of Programs with Resources

The graph-based semantics introduced in the previous chapter is often not informative enough, because it does not take into account whether two actions commute or not. In this chapter, we introduce truly concurrent models which incorporate this information. We begin by extending our programming languages with resources (Sect. 3.1) and restrict ourselves to conservative programs, in which resource consumption only depends on the current state (Sect. 3.2). We then generalize the semantics to asynchronous graphs, which explicitly describe the commutation of two actions (Sect. 3.3) and to precubical sets, which can more generally express the commutation of n actions (Sect. 3.4). Finally, links with other classical models for concurrency are mentioned (Sect. 3.5).

3.1 Modeling Resources in the Language

3.1.1 Taming Concurrency

The programming language introduced in the previous chapter is quite minimal and can of course be extended in many ways, in order to model more advanced features of programming languages (functions, objects, pattern matching, etc.). Since we focus here on concurrency aspects of programming languages, we will not detail those possible extensions. Starting from the next section, we however extend the language with a notion of resource, which will prove crucial in order to properly capture archetypal challenges in concurrency.

Concurrent programming unfortunately carries unique problems, not found in other modes of programming. When two threads access a shared resource, such as memory, the outcome is often unspecified. For instance, consider the program

$$x:=0;\ (x:=x+1\ \ ||\ \ x:=x+1) \tag{3.1}$$

© Springer International Publishing Switzerland 2016

L. Fajstrup et al., *Directed Algebraic Topology and Concurrency*,

DOI 10.1007/978-3-319-15398-8_3

in which two threads concurrently increment a variable x whose value is initially 0. The first intuition is that after the execution of the program the variable x should contain the value 2. However, because of the way threads and accesses to memory are implemented in practice, it happens that the variable might also contain 1, or even a completely unrelated value: in most programming languages concurrent access to shared memory is *unspecified*. In order to achieve reasonably predictable behavior when using shared memory, most operating systems provide a construction, called a *mutex* (short for *mut*ual *ex*clusion), which is a resource that can be held by at most one thread. Given such a mutex a, a thread can perform two operations on it [32]:

- *lock* the resource, which is modeled by the instruction P_a,
- *release* the resource, which is modeled by the instruction V_a.

The system guarantees that a mutex can be locked at most once: if a thread tries to lock a resource that has been previously locked by another thread, it remains frozen until the mutex is released (if multiple processes are frozen, only one of them is awaken when the mutex is released). In order to guarantee predictable behavior, the program (3.1) should thus be rewritten as

$$\texttt{x:=0;}\ (P_a;\, \texttt{x:=x+1;}\, V_a\ ||\ P_a;\, \texttt{x:=x+1;}\, V_a)$$

Another useful feature of mutexes is that they ensure that a sequence p of instructions is *atomic*, i.e., the sequence will never be interrupted: in a subprogram of the form $P_a;\, p;\, V_a$, we know that the instructions in p will not be interleaved with instructions from other subprograms running in parallel which are also using the locking and unlocking the mutex a in the same way. The portion of code between P_a and V_a is thus called a *blocking section*. The operations P and V are often called *synchronization primitives* because they help the programmer to regulate how threads will execute w.r.t. each other, and to make it easier to reason about concurrent programs. As an illustration of the use of atomic sequences, and more generally of the difficulty of verifying concurrent programs, consider the following program (on the left) launching two threads in parallel after an initialization phase:

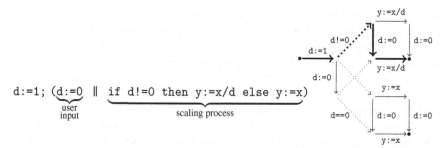

$$\texttt{d:=1;}\ (\underbrace{\texttt{d:=0}}_{\substack{\text{user}\\\text{input}}}\ ||\ \underbrace{\texttt{if d!=0 then y:=x/d else y:=x}}_{\text{scaling process}})$$

This program should be thought of as an image scaling program. Initially, the scaling factor d is set to 1. Then two processes are run in parallel: the first one takes the input of the user (taken here to be the scaling factor d to 0), and the second one takes care of the rescaling of the "image" (taken here to mean that the process takes a variable x thought of as the image, divides it by d, and stores the result in variable y). When the scaling factor d is zero, the scaling process does not operate on the image as scaling by zero is undefined. More realistically, the scaling process should be looping in order to regularly update the image, but adding a loop would not have much influence on the argument below. The transition graph of the above program is shown on the right above (the two endpoints on the right are the same, but have been drawn as separate vertices for clarity). At first it might seem that the scaling process is correct, in the sense that it will never lead to division by zero, since the division is performed only if the condition d!=0 is satisfied. However, because of the parallel construction, the user input process can be interleaved arbitrarily with the scaling process, and it might happen that the instruction d=0 is executed after the comparison d!=0 has been performed but before the execution of the instruction y = x/d (corresponding to the thick path in the above transition graph): in this case, a division by zero will be performed! In order to solve this problem, the programmer has to ensure that no instruction from another process will be executed between the comparison and the division, which can be achieved using mutexes as explained above. The program could thus be rewritten as

$$d:=1; \ (P_a; d:=0 : V_a \parallel P_a; \ (\text{if } d!=0 \text{ then } y:=x/d \text{ else } y:=x; V_a)$$

to avoid the above problem.

3.1.2 Extending the Language with Resources

We introduce a notion of "resource" in the language, which is more general than mutexes, in the sense that it can be specified to lock more than once. These resources are also sometimes called *counting semaphores* in the operating system literature.

Fix a set $\mathscr{R} = \{a, b, \ldots\}$ of *resources* together with a function $\kappa : \mathscr{R} \to \mathbb{N}$ associating to each resource a a *capacity* (or *arity*) $\kappa_a \in \mathbb{N}$, in particular a *mutex* is a resource whose capacity is 1. The syntax of programs given in Definition 2.1 is extended with two new families of constructions:

$$p \ ::= \ \ldots \ \mid \ P_a \ \mid \ V_a$$

where $a \in \mathscr{R}$ is a resource. If we write $P_{\mathscr{R}}$ (resp. $V_{\mathscr{R}}$) for the set $\{P_a \mid a \in \mathscr{R}\}$ (resp. $\{V_a \mid a \in \mathscr{R}\}$) of programs, the transition graph associated to a program is now labeled in $\mathscr{L} = \mathscr{C}_{\text{act}} \sqcup \mathscr{B} \sqcup P_{\mathscr{R}} \sqcup V_{\mathscr{R}}$, and is defined as in Definition 2.9 extended with the cases defining the graphs associated to the programs P_a and V_a, which are respectively

$$s_{P_a} \bullet \xrightarrow{\ P_a\ } \bullet t_{P_a} \qquad s_{V_a} \bullet \xrightarrow{\ V_a\ } \bullet t_{V_a}$$

The operational semantics introduced in Sect. 2.2.3 is then modified as follows. First, the set of states is now of the form $\Sigma = \mathbb{Z}^{Var} \times \mathbb{Z}^{\mathscr{R}}$, i.e., a state σ is now a pair $\sigma = (\sigma_v, \sigma_r)$ where $\sigma_v \in \mathbb{Z}^{Var}$ describes the contents of the variables as before, and $\sigma_r \in \mathbb{Z}^{\mathscr{R}}$ associates to each resource its *availability*. By abuse of notation, given a variable $x \in \textit{Var}$ (resp. a resource $a \in \mathscr{R}$), we often write $\sigma(x)$ instead of $\sigma_v(x)$ (resp. $\sigma(a)$ instead of $\sigma_r(a)$). The initial state σ_0 is such that for any variable $x \in \textit{Var}$, $\sigma_0(x) = 0$ and for any resource $a \in \mathscr{R}$, $\sigma_0(a) = \kappa_a$. The semantics for actions different from P_a and V_a is defined as before (as in Definition 2.18), except that variables are read from and written to σ_v, and that the σ_r component is left unmodified. For instance, the interpretation of variables and assignations is given by

$$\llbracket x \rrbracket_{\mathscr{A}}(\sigma) = (\sigma_v(x), \sigma_r) \qquad \llbracket x := a \rrbracket_{\mathscr{C}}(\sigma) = (\sigma_v \left[x \mapsto \llbracket a \rrbracket_{\mathscr{A}}(\sigma) \right], \sigma_r)$$

The action of the resource primitives is defined, for any resource $a \in \mathscr{R}$, by

$$\llbracket P_a; t \rrbracket(\sigma) = \llbracket t \rrbracket(\sigma_v, \delta_a^{-1}(\sigma_r)) \qquad \llbracket V_a; t \rrbracket(\sigma) = \llbracket t \rrbracket(\sigma_v, \delta_a^{+1}(\sigma_r))$$

where $\delta_a^{-1}(\sigma_r)$ (resp. $\delta_a^{+1}(\sigma_r)$) is the function which is the same as σ_r except on a, which has been decreased (resp. increased) by one; formally, $\delta_a^{-1}(\sigma_r)(a) = \sigma_r(a) - 1$, $\delta_a^{-1}(\sigma_r)(b) = \sigma_r(b)$ for $b \neq a$, and similarly for $\delta_a^{+1}(\sigma_r)$. The notion of *validity* extends Definition 2.19 by allowing words in $(\mathscr{C}_{\text{act}} \sqcup \mathscr{B} \sqcup P_{\mathscr{R}} \sqcup V_{\mathscr{R}})^*$ of the form

- $t.P_a$ with $a \in \mathscr{R}$ and $\llbracket t \rrbracket(\sigma_0)(a) > 0$
- $t.V_a$ with $a \in \mathscr{R}$ and $\llbracket t \rrbracket(\sigma_0)(a) < \kappa_a$

which expresses the fact that in a valid execution trace, the locked resources have to be available and one cannot add more instances of a resource than its capacity.

Remark 3.1 We could have implemented the primitives P_a and V_a within the language, but this is far from being easy [32]. In particular, to implement P_a we have to ensure that $a > 0$, and then decrement a without some other thread decrementing a in between. Similarly, general resources with arbitrary capacity can be implemented from those with capacity one, i.e., mutexes. The point of having general resources as basic constructs in the language allows us to enforce a certain discipline on their usage, as we will see in the next section.

Remark 3.2 The assumption that $\sigma_0(a) = \kappa_a$ means that resources are available to the maximum of their capacity at the beginning of the program. This assumption is not restrictive: studying a program p in which the initial state σ_0 is such that $0 \le \sigma_0(a) < \kappa_a$, is equivalent to study the program $P_a; P_a; \ldots; P_a; p$, with $\kappa_a - \sigma_0(a)$ occurrences of P_a at the beginning, in the initial state σ_0' defined as σ_0 except $\sigma_0'(a) = \kappa_a$.

Remark 3.3 The constructions provided in our programming language are for instance close to those provided by Java [61] where threads are defined and executed as objects of a `Thread` class (or implementing the `Runnable` interface). Moreover, the operational semantics of both languages are close, although the compliance with interleaving semantics is not complete. In practice, on modern platforms, the code is not always executed in the order it was written, and sequential consistency (see Remark 2.17) or even linearizability [91] is not ensured, unless some particular conditions are met (for instance, the memory locations on which writes are done are declared `volatile`). For the purpose of this book though, the more idealistic semantics we are considering is enough. It covers well the synchronization aspects met in Java, which has, as in the POSIX standard [79] and implementations, semaphores (including mutexes), ensuring mutual exclusion properties, and monitors [32]. This allows for implementing weaker synchronization primitives such as counting semaphores [31] as presented in this book.

Remark 3.4 The resources we have chosen to model are close to what is usually called *semaphores*, of which mutexes are a particular case. There are many other synchronization primitives that can be used (e.g., monitors, barriers) to ease the implementation of some idiomatic structures (e.g., queues, message-passing concurrency). They could be studied in a similar way, either by modifying accordingly the semantics or by implementing them with semaphores.

We will provide examples of programs with resources in the next sections, and many other can be found in [33].

3.2 State Spaces for Conservative Resources

3.2.1 Conservative Programs

In order to study the resource consumption of a program, we introduce the following notion which expresses the overall effect of a program on the resources.

Definition 3.5 Given a program p, its *resource consumption* $\Delta(p) : \mathscr{R} \to \mathbb{Z}$ gives, for each resource a, the number $\Delta(p)(a)$ of resources a it has taken or released (depending on whether this number is negative or positive), i.e., the difference between the number of V_a instructions and the number of P_a instructions encountered in an execution of p. It is defined by induction on p by

$$
\begin{array}{llll}
\Delta(A) & = & 0 & \Delta(\texttt{skip}) & = & 0 \\
\Delta(P_a) & = & -\delta_a & \Delta(V_a) & = & \delta_a \\
\Delta(p;q) & = & \Delta(p)+\Delta(q) & \Delta(p\,\|\,q) & = & \Delta(p)+\Delta(q) \\
\Delta(\texttt{if } b \texttt{ then } p \texttt{ else } q) & = & \Delta(p) & \text{whenever } \Delta(p)=\Delta(q) \\
\Delta(\texttt{while } b \texttt{ do } p) & = & 0 & \text{whenever } \Delta(p)=0
\end{array}
$$

where A is an arbitrary action. Above, 0 denotes the constant function whose image is 0, the addition of two functions is the pointwise addition and δ_a denotes the function such that $\delta_a(a) = 1$ and $\delta_a(b) = 0$ for any $b \neq a$. Notice that the function is only partially defined because of the side conditions in the cases of branching and loop.

Proposition 3.6 *Given a program p such that $\Delta(p)$ is defined, for any path $t : s_p \twoheadrightarrow t_p$ in G_p and resource $a \in \mathscr{R}$, we have $[\![t]\!](\sigma_0)(a) = \kappa_a + \Delta(p)(a)$.*

The programs for which the resource consumption is well-defined can be character-ized as those which are conservative, in the following sense.

Definition 3.7 A program p is **conservative** (or *well-bracketed*) w.r.t. resources when for any state $\sigma \in \Sigma$, any pair of paths $t, u : x \twoheadrightarrow y$ in G_p, with same source and same target, and any resource $a \in \mathcal{R}$, we have $[\![t]\!](\sigma)(a) = [\![u]\!](\sigma)(a)$.

Proposition 3.8 *A program p is conservative if and only if $\Delta(p)$ is well-defined.*

Example 3.9 A program of the form while b P_a is not conservative: indeed, the two paths $t, u : s_p \twoheadrightarrow t_p$ which are respectively labeled by $b.P_a$ and $b.P_a.b.P_a$ satisfy $[\![t]\!](\sigma_0)(a) = \kappa_a - 1$ and $[\![u]\!](\sigma_0)(a) = \kappa_a - 2$.

Remark 3.10 From the above proposition, determining whether a program p is con-servative or not can be done by checking whether $\Delta(p)$ is defined or not, and this can be done in linear time w.r.t. the size of p by directly implementing Definition 3.5

Definition 3.11 Given a conservative program p, for every vertex x in G_p, we define the *resource potential* $r(x) : \mathcal{R} \to \mathbb{Z}$ at x by $r(x)(a) = [\![t]\!](\sigma_0)(a)$ for any resource $a \in \mathcal{R}$, where $t : s_p \twoheadrightarrow x$ is a path in G_p. The natural number $r(x)(a)$ is called the *residual capacity* of the resource a at position x.

Notice that the above definition is well-defined: for any vertex x there is a path $t : s_p \twoheadrightarrow x$ by Lemma 2.12, and the potential does not depend on the choice of the path t since p is conservative. The following proposition shows that resource consumption along any path can be computed from the resource potential:

Proposition 3.12 *For any path $t : x \twoheadrightarrow y$ in G_p, state $\sigma \in \Sigma$ and resource $a \in \mathcal{R}$, we have $[\![t]\!](\sigma)(a) = \sigma(a) + r(y)(a) - r(x)(a)$.*

Remark 3.13 The terminology *conservative* comes from an analogy with physics: a force is conservative when its work along a path only depends on the endpoints of the latter. In this case, the force derives from a potential, which—following our analogy—corresponds to the resources in a position.

The above proposition shows that, in a conservative program p, the resource consumption of any potential execution path $t : s_p \twoheadrightarrow x$ only depends on the target vertex x and is given by $r(x)$. In particular, suppose that a path t goes through a vertex z which is such that $r(z)(a) < 0$. Since $r(s_p)(a) \geq 0$, and the operation P_a is the only one which can decrease resource availability of a and it only decreases it by one, we know that the path t is of the form

$$t = s_p \overset{u}{\twoheadrightarrow} x \overset{P_a}{\to} y \overset{v}{\twoheadrightarrow} z \overset{w}{\twoheadrightarrow} z'$$

with $r(x)(a) = 0$ and $r(y)(a) = -1$. Therefore, the path t is not valid (in the sense of Sect. 3.1.2) since it contains a prefix $u.P_a$ with $[\![u]\!](\sigma_0)(a) = r(x)(a) = 0$ and a valid path should satisfy $[\![u]\!](\sigma_0)(a) > 0$. We have just shown that no valid path goes through a state x with $r(x)(a) < 0$ for some resource $a \in \mathcal{R}$. A similar reasoning can be held for paths going through a vertex x with $r(x)(a) > \kappa_a$.

Definition 3.14 A vertex is *valid* (and *forbidden* otherwise) when for every resource $a \in \mathscr{R}$, we have $0 \leq r(x)(a) \leq \kappa_a$.

Proposition 3.15 *For any execution trace* $t : s_p \twoheadrightarrow x$ *reaching a vertex* x, *the vertex* x *is valid.*

The previous proposition can be rephrased more concisely by saying that vertices which are not valid are unreachable (in the sense of Definition 2.28).

3.2.2 Transition Graphs for Conservative Programs

In the following, we will suppose that all the programs we consider are conservative (this is not very restrictive in practice, and anyway the condition can easily be checked by Remark 3.10). As we have seen in the previous section, we know that no valid path can go through a vertex which is not valid. We can therefore restrict the transition graph to vertices which are valid without removing (valid) execution traces. By restricting, we mean the following.

Definition 3.16 The **pruned transition graph** $\check{G}_p$ is obtained from the transition graph by restricting to valid vertices (in the sense of Definition 3.14), keeping the terminal position. Formally, if we write V' for the set of valid vertices, we have

$$\check{G}_p = \begin{cases} G_p\big|_{V'} & \text{if } t_p \in V' \\ G_p\big|_{V'} \sqcup t_p & \text{otherwise} \end{cases}$$

where t_p denotes the graph containing t_p as only vertex and no edge.

Remark 3.17 The initial position s_p of a program is always valid, which explains why we do not have special case similar to the one for the end position t_p in the preceding definition.

Lemma 3.18 *The inclusion* $\check{G}_p \hookrightarrow G_p$ *induces a bijection between valid paths from the initial vertex (i.e., execution traces) in the two graphs.*

We saw in Lemma 2.12 that every vertex can be reached from the initial vertex s_p and can reach the final vertex t_p in the transition graph G_p of a program p. Since we have removed vertices during the pruning, this is not necessarily true anymore in $\check{G}_p$. This is quite interesting since it enables us to discover some of the problematic positions described in Definition 2.28.

Proposition 3.19 *Given a conservative program* p, *the following holds.*

- *A position* x *in* $\check{G}_p$ *such that there is no path from the initial vertex* s_p *to* x *is unreachable (positions which are not reachable in* $\check{G}_p$ *are unreachable).*

- *A position x in $\check{G}_p$, which is different from t_p, reachable by an execution trace, and such that there is no edge with x as source in $\check{G}_p$, is a deadlock.*
- *A position x in $\check{G}_p$, such that there exists an execution trace reaching a deadlock and there is a prefix of this trace reaching x, is unsafe.*
- *A position x in $\check{G}_p$, which is reachable by an execution trace, and such that there is no path from x to t_p, is doomed.*

Remark 3.20 Notice that there is a subtle difference in the above proposition between a position x of $\check{G}_p$ which is "not reachable" from s_p, which means that there is no path from s_p to x in the graph $\check{G}_p$, and "unreachable," which will consistently be taken in the sense of Definition 2.28, meaning that there is no execution trace reaching x, i.e., no *valid* path from s_p to x in $\check{G}_p$.

The previous proposition is quite useful, since it can be directly used to implement a simple verification algorithm. For the properties presented in Sect. 2.3.2.

Algorithm 3.21 Undesirable positions, in the sense of Definition 2.28, can be discovered from the pruned transition graph of a conservative program as follows:

- a position which is not reachable from the beginning position is unreachable,
- a position different from the end position and from which there is no transition is a *potential deadlock*: it is either a deadlock or unreachable,
- a position from which there is a path to a deadlock is *potentially unsafe*: it is either unsafe or unreachable,
- a position from which there is no path to the end position is *potentially doomed*: it is either doomed or unreachable.

Above, the positions only "potentially" have these properties because they might not be reachable, we will generally omit mentioning it in the following. The algorithm is not *complete*: it only finds the positions satisfying the above properties (such as potential deadlocks) because of the "structure of the program," i.e., the way synchronization primitives are used, but does not consider the values manipulated by the program, which explains why it cannot find all such positions. We first provide some examples, and discuss the reasons preventing the algorithm from being complete in Remark 3.25.

Example 3.22 (*Swiss flag*) Consider the following program p:

$$P_a; P_b; V_b; V_a \quad || \quad P_b; P_a; V_a; V_b$$

with $a, b \in \mathscr{R}$ mutexes ($\kappa_a = \kappa_b = 1$). We have drawn the transition graph G_p on the left, and the pruned transition graph $\check{G}_p$ on the right:

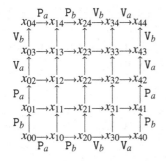

The beginning and end vertex are respectively $s_p = x_{00}$ and $t_p = x_{44}$. We have only shown the labels for the exterior edges, the labels for other edges can be deduced using the convention that two edges drawn in parallel have the same label, e.g., the edge $x_{22} \rightarrow x_{32}$ is labeled by V_b. This example is often called the *Swiss flag* because of the shape of the pruned transition graph.

Since both a and b are mutexes, the resource potential at the beginning position is $r(x_{00})(a) = r(x_{00})(b) = 1$. Because the transition $x_{00} \rightarrow x_{10}$ is labeled by P_a, and the action of P_a is to decrement the number of resources a (see Sect. 3.1.2), we have $r(x_{10})(a) = 0$ and $r(x_{10})(b) = 1$. By reasoning similarly on transitions $x_{10} \rightarrow x_{20}$ and $x_{20} \rightarrow x_{21}$, we have $r(x_{20})(a) = r(x_{20})(b) = 0$, as well as $r(x_{21})(a) = 0$ and $r(x_{21})(b) = -1$. Therefore, the state x_{21} is not valid. It can be shown in the same way that the states x_{12}, x_{22}, x_{32} and x_{23} are not valid either, and that all other states are valid. The pruned transition graph is then obtained from the transition graph by removing invalid vertices, as well as edges having those vertices as source or target.

Notice that in the pruned transition graph the vertex x_{11} is a deadlock since it is distinct from the end vertex x_{44} and no transition originates from it: it corresponds to the situation where the first process has taken the resource a and is waiting for the resource b, while the other process has taken the resource b and is waiting for the resource a. The vertex x_{33} is unreachable since it is not the beginning vertex x_{00} and it is not the target of any transition.

Example 3.23 (Dining philosophers) We recast here the well-known example of the dining philosophers due to Dijkstra, as reformulated by Hoare [93]. Consider n Chinese philosophers seated together around a round table and ready to eat a meal already served. Between each two philosophers there is a chopstick. A philosopher has to take both the chopstick on his left and on his right, eat, and then put them back. For instance, if we suppose that the philosophers start by taking the chopstick on their left before the one on the right, the situation can be modeled as follows. The chopsticks are represented by n mutexes a_i (with $\kappa_{a_i} = 1$) and each philosopher is modeled by a processes p_i of the form

$$p_i = P_{a_i}; P_{a_{i+1}}; A; V_{a_i}; V_{a_{i+1}}$$

where the indices i, above, are to be considered modulo n, so that they satisfy $1 \leq i \leq n$. The two P actions correspond to taking the two chopsticks, the action A

corresponds to eating, and the two V actions correspond to putting the chopsticks back on the table. The general situation is modeled by the process $p = p_1 \parallel p_2 \parallel \cdots \parallel p_n$ corresponding to the n philosophers around the table. For instance, with two and three processes, the pruned transition graphs are respectively

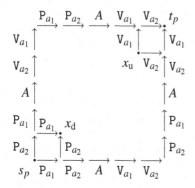
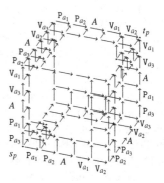

Notice that with two philosophers the position x_d is a deadlock. This corresponds to the situation where each philosopher has taken his left chopstick and is waiting for the other philosopher to release the chopstick he has taken: they are stuck indefinitely in this situation and will never eat. Dually, the position x_u is unreachable. This corresponds to the situation where both philosophers are ready to release their right chopstick which means that both have already taken their left chopstick (because they take the left one before the right one and release the right one before the left one in our example) and therefore each chopstick would have been taken by both philosophers, which is impossible. Similar positions can be found in the example with three philosophers. Notice that if one of the philosophers is willing to take his right chopstick before his left one then the deadlock vanishes. For instance, if we replace p_1 by $P_{a_2}; P_{a_1}; A; V_{a_2}; V_{a_1}$ while leaving the other unchanged, with two and three processes we get the following pruned transition graphs without deadlocks or unreachable positions:

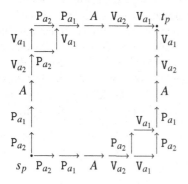
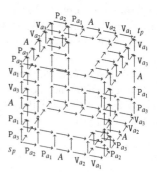

Example 3.24 In the following, we will sometimes write `deadlock` for the quintessential deadlock program:

$$deadlock = P_a; V_a$$

with $\kappa_a = 0$, whose transition graph is shown on the left:

$s_{deadlock}$ •   • $t_{deadlock}$ $s_{deadlock'}$ •$\xrightarrow{\ P_a\ }$•   •$\xrightarrow{\ V_a\ }$• $t_{deadlock'}$

Here, the position s_{deadlock} is clearly a deadlock and the position t_{deadlock} is unreachable. Notice that the even simpler program P_a exhibits the same situation (with the end position being not valid, illustrating why we might have to add it again in Definition 3.16). However, the above program will be simpler to use in other programs while having those programs be conservative (as in Remark 3.25). If the reader is worried about using resources of capacity 0 and is more assured with mutexes, the following program $deadlock' = P_a; P_a; V_a; V_a$ with $\kappa_a = 1$, whose transition graph is shown on the right, can generally be used instead in the examples.

Remark 3.25 It has already been noticed that Algorithm 3.21 only allows us to discover some undesirable positions, but not all of them since we do not consider the semantics of instructions. For instance, consider the program

$$p = \text{if false then skip else skip}$$

whose pruned transition graph is shown on the left

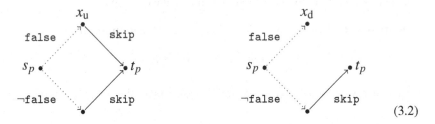

$$(3.2)$$

The position x_{u} is unreachable because the boolean condition `false` is never true, but not discovered by the algorithm. The following variant is also interesting:

$$p = \text{if false then } deadlock \text{ else skip}$$

Its pruned transition graph is shown on the right of (3.2). The position x_{d} is discovered as a potential deadlock, but it is not a deadlock in the sense of Definition 2.28 because it is unreachable for similar reasons as above (however, because of the specific programming language we chose, it can be shown that all deadlocks are discovered by the algorithm). Finally, consider the following program p whose pruned transition graph is shown on the right:

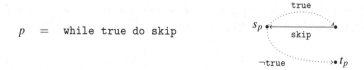

$$p \quad = \quad \text{while true do skip}$$

The end position t_p is unreachable because the condition $\neg \text{true}$ is never true. However, it is not discovered by the algorithm. This example is quite similar to the previous one, but uses a while loop instead of a conditional branching. The above examples make it clear that we could prune more by also removing vertices which cannot be reached because branching conditions cannot be satisfied. However, this would render pruning undecidable (even though we could use some heuristics or static analysis to remove some of them), and moreover, as mentioned earlier, the aim of this book is to focus on features of programming languages which are specific to concurrency.

3.3 Asynchronous Semantics

3.3.1 Toward True Concurrency

As explained in Sect. 2.3, the number of paths to verify can be exponential in the number of threads. In order to lower this number, we have to take commutations of actions into account: two actions commute when the effect of their action on any state does not depend on the order in which they are executed.

Definition 3.26 Two actions $A, B \in \mathscr{C}_{\text{act}} \sqcup \mathrm{P}_{\mathscr{R}} \sqcup \mathrm{V}_{\mathscr{R}}$ **commute** when

$$[\![B]\!] \circ [\![A]\!] = [\![A]\!] \circ [\![B]\!]$$

Put another way, for any transition graph containing a subgraph of the form

$$
\begin{array}{ccc}
y_2 & \xrightarrow{\ A\ } & z \\
{\scriptstyle B}\Big\uparrow & & \Big\uparrow{\scriptstyle B} \\
x & \xrightarrow[\ A\ ]{} & y_1
\end{array}
\tag{3.3}
$$

the semantics of the two paths from x to z are the same.

Example 3.27 The two actions A and B, which are respectively x:=2*x and x:=x+1, do not commute because we have $1 = [\![B]\!] \circ [\![A]\!](\sigma_0) \neq [\![A]\!] \circ [\![B]\!](\sigma_0) = 2$ (we recall that we have $\sigma_0(x) = 0$).

Example 3.28 The two actions $x:=2*x$ and $y:=y+1$ are easily checked to commute.

Remark 3.29 Since the only paths we are interested in when verifying programs are the execution traces, the notion of commutation between actions which is really interesting for our purposes is the following one: a square of the form (3.3) *commutes* when, for every path $t : s_p \twoheadrightarrow x$, we have $[\![B]\!] \circ [\![A]\!] \circ [\![t]\!](\sigma_0) = [\![A]\!] \circ [\![B]\!] \circ [\![t]\!](\sigma_0)$. This notion of commutation is impossible to compute in practice, i.e., it is undecidable whether four transitions are commuting, in contrast to the notion of Definition 3.26. Moreover, actions commuting in the sense of Definition 3.26 are commuting in the sense of this remark, which makes it a suitable over-approximation.

The commutation given in Example 3.28 is an instance of a more general fact. Given an action A, we write $\mathrm{FV}(A)$ for the set of (free) variables occurring in it. For instance, $\mathrm{FV}(x:=2*y) = \{x, y\}$.

Lemma 3.30 *Any two actions A and B such that $\mathrm{FV}(A) \cap \mathrm{FV}(B) = \emptyset$ commute.*

Remark 3.31 The previous lemma could be refined by distinguishing between variables which are used for reading and those used for writing: if a common variable is used only for reading, the two actions still commute. For instance, the actions $y:=2*x$ and $z:=x+5$ commute, even though they share x as free variable.

3.3.2 Asynchronous Semantics

In this section, we provide a semantics for concurrent programs using asynchronous graphs, which are graphs equipped with a notion of equivalence between paths, used to keep track of the commutation between actions described in the previous section. This model is introduced here mostly for didactic and historic purposes: it can be considered as a bridge between the sequential models based on graphs used up to now, and the precubical models which are richer—and thus a bit more difficult to grasp at first—since they can encode commutations between any n-tuple of actions, as opposed to only pairs of actions.

Definition 3.32 A labeled **asynchronous graph** (G, I) consists of a labeled graph G together with a set of *independence tiles* I consisting of pairs $(e_1.e_2', e_2.e_1')$ of paths of length 2 with the same source and the same target

$$
\begin{array}{ccc}
y_2 & \xrightarrow{\;e_1'\;} & z \\
{\scriptstyle e_2}\Big\uparrow & \sim & \Big\uparrow{\scriptstyle e_2'} \\
x & \xrightarrow[\;e_1\;]{} & y_1
\end{array}
$$

$$(3.4)$$

such that $\ell(e_1) = \ell(e_1')$ and $\ell(e_2) = \ell(e_2')$, and I forms a relation which is symmetric. A *morphism* f between two labeled asynchronous graphs (G, I) and (G', I') is a morphism $f : G \to G'$ between the underlying labeled graphs, such that for every pair of paths $(e_1.e_2', e_2.e_1')$ in I, their image $(f(e_1).f(e_2'), f(e_2).f(e_1'))$ is in I'.

We often use the symbol $\sim$, as in (3.4), to indicate the elements of I when drawing an asynchronous graph.

An asynchronous graph can be seen abstractly as a two-dimensional space: the vertices can be thought as points, the edges as generators for paths, and the tiles as (generators for) two-dimensional surfaces between paths. This is what is suggested by the diagram (3.4): since there is a surface between the paths $e_1 . e_2'$ and $e_2 . e_1'$, we can consider that we have the possibility of "continuously" deforming the first path into the second. It thus seems natural to call the relation introduced in the next definition dihomotopy, since it is a combinatorial version of homotopy between directed paths. This point of view will be extensively explained and developed in the rest of this chapter and in the following one, see Sect. 4.2 in particular.

Definition 3.33 Given an asynchronous graph (G, I), the **dihomotopy** relation on paths, denoted by $\sim$, is the smallest equivalence relation on paths of G which contains I and is a congruence w.r.t. concatenation:

- given a pair of paths $(s_1, s_2) \in I$ we have $s_1 \sim s_2$,
- and for all paths $s_1, s_2 : x \twoheadrightarrow y$, $u : x' \twoheadrightarrow x$ and $v : y \twoheadrightarrow y'$, we have

$$s_1 \sim s_2 \qquad \text{implies} \qquad u . s_1 . v \sim u . s_2 . v$$

One can easily show that two paths which are dihomotopic necessarily have the same source and the same target. Moreover, since the dihomotopy relation is a congruence by definition, the following category can be associated to any asynchronous graph; it will play a major role in Chap. 4.

Definition 3.34 Given an asynchronous graph (G, I), its **fundamental category** $\vec{\Pi}_1(G, I)$ is the category whose objects are the vertices of the graph and morphisms from x to y are equivalence classes of paths from x to y by the dihomotopy relation. In this case, any category isomorphic to $\vec{\Pi}_1(G, I)$ is said to be *presented* by the asynchronous graph (G, I).

The operations used in Sect. 2.2.2 can be extended to asynchronous graphs in order to give a semantics within asynchronous graphs. Namely, we define the following operations, extending those of Definition 2.6 on graphs.

- The *disjoint union* $(G_1, I_1) \sqcup (G_2, I_2)$ of two asynchronous graphs (G_1, I_1) and (G_2, I_2) is the asynchronous graph $(G_1 \sqcup G_2, I_1 \sqcup I_2)$.
- The *tensor product* $(G_1, I_1) \otimes (G_2, I_2)$ of two asynchronous graphs (G_1, I_1) and (G_2, I_2) is the asynchronous graph $(G_1 \otimes G_2, I)$ where a tile in I is of one of the following forms:

1. $((e_1, x_2).(e'_2, x_2), (e_2, x_2).(e'_1, x_2))$ with $(e_1.e'_2, e_2.e'_1) \in I_1$ and $x_2 \in V_2$
2. $((x_1, e_1).(x_1, e'_2), (x_1, e_2).(x_1, e'_1))$ with $x_1 \in V_1$ and $(e_1.e'_2, e_2.e'_1) \in I_2$
3. $((e_1, x_2).(y_1, e_2), (x_1, e_2).(e_1, y_2))$ with $e_1 : x_1 \to y_1$ edge of G_1 and $e_2 : x_2 \to y_2$ edge of G_2, or symmetrically.

$$
\begin{array}{ccc}
(e'_1,x_2) & & \\
(y_2,x_2) \longrightarrow (z,x_2) & (x_1,y_2) \overset{(x_1,e'_1)}{\longrightarrow} (x_1,z) & (x_1,y_2) \overset{(e_1,y_2)}{\longrightarrow} (x_2,y_2) \\
(e_2,x_2) \;\sim\; (e'_2,x_2) & (x_1,e_2) \;\sim\; (x_1,e'_2) & (x_1,e_2) \;\sim\; (y_1,e_2) \\
(x,x_2) \longrightarrow (y_1,x_2) & (x_1,x) \longrightarrow (x_1,y_1) & (x_1,x_2) \longrightarrow (y_1,x_2) \\
(e_1,x_2) & (x_1,e_1) & (e_1,x_2)
\end{array}
$$

An edge in $G_1 \otimes G_2$ comes from either G_1 or G_2. Two such edges thus commute when they either come from different graphs, or they come from the same graph and they already commuted in this graph.

- The *quotient* $(G, I)[x = y]$ of an asynchronous graph (G, I) by identifying two vertices x and y is $(G[x = y], I)$.
- The *restriction* $(G, I)|_{V'}$ of an asynchronous graph (G, I) to a subset V' of the vertices is the asynchronous graph $(G|_{V'}, I')$ where I' is the subset of I of those pairs of paths whose constituent edges have both their source and their target in V'.

Using these operations it is easy to define a model for programs:

Definition 3.35 The *asynchronous transition graph* G_p associated to a program p is then defined as in Definition 2.9, using the above operations on asynchronous graphs. The *pruned asynchronous transition graph* $\check{G}_p$, also called the **asynchronous semantics** of p, is then obtained by restricting to valid vertices as in Definition 3.16.

Example 3.36 Consider a program of the form $p = A \parallel B$ where A and B are arbitrary actions. We have $\check{G}_p = G_p = G_A \otimes G_B$ and its asynchronous semantics is shown on the left

The tile was introduced here by the tensor operation which declares as independent two actions coming from two threads in parallel, such as A and B: every square with sides of length 2, as in (3.4), whose vertical transitions and horizontal transitions come from actions in two distinct threads has its sides related by an independence tile. Because of this, the asynchronous semantics associated with the Swiss flag (Example 3.22, which is detailed in Example 3.40) dining philosophers (Example 3.23) and many other previous example programs are obtained from their respective pruned transition graphs by adding a tile for every square with sides of length 2. Not every such square is filled by a tile though: we have for instance shown the asynchronous transition graph of $p = $ if b then A else B on the right.

Remark 3.37 In the previous example, it should be noticed that in the asynchronous semantics of $A \parallel B$, the paths $A . B$ and $B . A$ are always dihomotopic, even though the actions A and B might not commute in the sense of Definition 3.26 (for instance when A and B are respectively $x := 0$ and $x := 1$). The cases where the two notions are related are of course the desirable ones and those will be studied in next Sect. 3.3.3.

Example 3.38 The asynchronous semantics $\check{G}_p$ of the following program

$$p = (x := 1 \parallel y := 2); \; (P_a; z := x; V_a \parallel P_a; z := y; V_a)$$

where a is a mutex, is

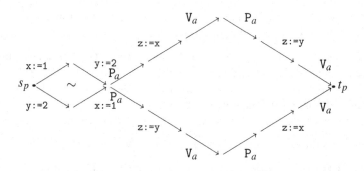

which, as expected, encodes the fact that the actions $x := 1$ and $y := 2$ commute, but the actions $z := x$ and $z := y$ do not.

Example 3.39 The previous example illustrates the meaning of holes w.r.t. the semantics of programs. However, we will often be interested in shorter examples, not involving manipulation of data, but still exhibiting a behavior which is interesting from the point of view of concurrency. For instance, the conservative programs $P_a; V_a \parallel P_b; V_b$ and $P_a; V_a \parallel P_a; V_a$, respectively have the following asynchronous semantics:

$$
\begin{array}{ccc}
\begin{array}{ccc}
 & P_a & V_a \; t_p \\
V_b \uparrow & \sim \uparrow & \sim \uparrow V_b \\
P_b \uparrow & \sim \uparrow & \sim \uparrow P_b \\
s_p & P_a & V_a
\end{array}
&
\begin{array}{ccc}
 & P_a & V_a \; t_p \\
V_a \uparrow & & \uparrow V_a \\
P_a \uparrow & & \uparrow P_a \\
s_p & P_a & V_a
\end{array}
\end{array}
$$

Example 3.40 The asynchronous semantics of the Swiss flag (Example 3.22) has the pruned transition graph given in Example 3.22 as underlying graph and I is the symmetric closure of

$$\left\{ \left(x_{00} \xrightarrow{P_a} x_{10} \xrightarrow{P_b} x_{11}, x_{00} \xrightarrow{P_b} x_{01} \xrightarrow{P_a} x_{11} \right), \left(x_{03} \xrightarrow{P_a} x_{13} \xrightarrow{V_b} x_{14}, x_{03} \xrightarrow{V_b} x_{04} \xrightarrow{P_a} x_{14} \right), \right.$$

$$\left. \left(x_{30} \xrightarrow{V_a} x_{40} \xrightarrow{P_b} x_{41}, x_{30} \xrightarrow{P_b} x_{31} \xrightarrow{V_a} x_{41} \right), \left(x_{33} \xrightarrow{V_a} x_{43} \xrightarrow{V_b} x_{44}, x_{33} \xrightarrow{V_b} x_{34} \xrightarrow{V_a} x_{44} \right) \right\}$$

In the following, in order to tackle the problem of state space explosion explained in Sect. 2.3.2, the general idea will be to consider execution traces up to dihomotopy, i.e., up to commutation of independent actions, in order to reduce the number of schedulings to consider. The following example illustrates the fact that this reduction can be quite important in some cases.

Example 3.41 Consider Example 3.23, that is, the dining philosophers. Below are, for small values of n (the number of philosophers), the number of states, deadlocks, maximal traces (including those leading to the deadlock), total traces (going from the beginning to the end state), and maximal and total traces up to dihomotopy of the philosophers problem:

n	States	Deadlocks	Maximal traces	Total traces	Maximal traces up to dihom.	Total traces up to dihom.
2	21	1	4	2	3	2
3	99	1	912	906	7	6
4	465	1	648348	648324	15	14

For n philosophers, there are more than 2^{2n} states, and more than $2^{(n-1)^2}$ total traces, and hence the state space and the path space are growing exponentially in the number of philosophers. In comparison, there are only $2^n - 1$ classes of maximal traces up to dihomotopy, among which $2^n - 2$ are total (and one is leading to a deadlock), which is much less than the number of traces without the quotient. This idea will be the starting point of the developments from Chap. 4 on, with the notion of trace up to (di-)homotopy. A more trivial illustration of this phenomenon would be n processes $p_i = \mathrm{P}_{a_i}; \mathrm{V}_{a_i}$, locking distinct resources, running in parallel. The number of states is 3^n, and the number of maximal traces (equal to total traces) is bigger than $(n + 1)!$. But there is just one total execution trace up to dihomotopy, and only one interesting state (or "component," see Chap. 6).

3.3.3 Coherent Programs

In previous examples, it was noticed that the dihomotopy in the asynchronous graph semantics was the "expected one," i.e., two paths are homotopic when one can be transformed into the other by permuting commuting actions. It can however be noticed that nothing guarantees that the dihomotopy in the asynchronous graph model (Definition 3.33) coincides with the semantic one (Definition 3.26), as noticed in Remark 3.37. Programs for which it is the case are called coherent: in such a program, two dihomotopic paths have the same semantics.

Definition 3.42 A conservative program p is **coherent** when for every two dihomotopic paths $t, u : x \twoheadrightarrow y$ in the asynchronous semantics $\check{G}_p$, t is an execution trace if and only if u is an execution trace, and in this case we have $[\![t]\!] = [\![u]\!]$.

Another way to state the coherence property for a program p is that the semantics is well-defined on morphisms of the fundamental category $\vec{\Pi}_1(\check{G}_p)$.

Remark 3.43 Given a conservative program p, the dihomotopy in the asynchronous semantics $(\check{G}_p, \check{I}_p)$ is generated by the independence relation $\check{I}_p$, see Definition 3.33. The program p is therefore coherent if and only if for every pair of paths $(t, u) \in I$, we have $[\![t]\!] = [\![u]\!]$ and one is an execution trace if and only if the other is. The latter condition being the one used in practice.

Example 3.44 The program

$$p = (\texttt{x:=0} \quad \| \quad \texttt{x:=1}); \ \text{if} \ \texttt{x==0} \ \text{then} \ \textit{deadlock} \ \text{else} \ \text{skip}$$

whose asynchronous semantics is

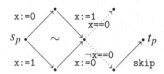

is *not* coherent: the two paths $t = \texttt{x:=0}.\texttt{x:=1}$ and $u = \texttt{x:=1}.\texttt{x:=0}$ are related by an independence tile, and are thus dihomotopic, but of course their semantics are not the same since, for any state σ, we have $[\![t]\!](\sigma)(\texttt{x}) = 1 \neq 0 = [\![u]\!](\sigma)(\texttt{x})$. And in fact, here, the order in which the two instructions are executed really matters since the program will either reach the end position t_p or be stuck in a deadlock depending on the scheduling. Notice however that if we had "protected" the parallel accesses to the variable $\texttt{x}$, the program would be coherent. Namely, the following variant of the program is coherent, where a is a mutex:

$$(\text{P}_a;\texttt{x:=0};\text{V}_a\|\text{P}_a;\texttt{x:=1};\text{V}_a); \ \text{if} \ \texttt{x==0} \ \text{then} \ \textit{deadlock} \ \text{else} \ \text{skip}$$

Notice that in the semantics dihomotopic paths are generated by the "$\|$" instruction, which gets interpreted as a tensor of asynchronous graphs. For sequential programs (without "$\|$"), the dihomotopy relation is reduced to equality ($t \sim u$ implies $t = u$). Coherence is thus immediate in this case.

Lemma 3.45 *Every sequential program is coherent.*

Since in a coherent program we know that two dihomotopic execution traces have the same effect on the state, in order that the executions of the program are valid, it is enough to check one representative in each dihomotopy class of execution traces (and not all the representatives). This can be efficiently exploited in order to reduce the space of traces to explore during verification, and gives results which are comparable to *partial-order reduction* techniques [60, 159] which are for instance implemented in the SPIN tool [94], as detailed in [71]. Even if execution traces are considered up to dihomotopy, a program can still generate an exponential number of those, thus showing the intrinsic difficulty of verifying concurrent programs, see for instance the n dining philosophers problem, Example 3.23. Similarly, the example given at the beginning of Sect. 2.3.2 can easily be adapted: the program $p = q \parallel \ldots \parallel q$ with n copies of $q = P_a; V_a$, where a is a mutex, has $n!$ dihomotopy classes of execution traces.

In the following, we are going to suppose that all the programs we manipulate are coherent. The fact that this assumption is satisfied is left to the programmer, as in the POSIX philosophy: the main way of ensuring coherence is to use mutexes as in the previous example. This is necessary because, in general, the property of being coherent for a program is undecidable. Notice that instead of supposing that coherence is ensured by the programmer, we could have as well modified the asynchronous semantics in order to remove independence tiles from $\check{G}_p$ of the form of the left below such that for every execution trace $t : s_p \twoheadrightarrow x$ we have $[\![t . A . B]\!](\sigma_0) = [\![t . B . A]\!](\sigma_0)$, as shown for instance in the examples in the middle and the right:

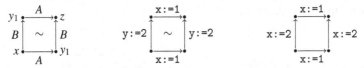

While this solution would be satisfactory from a theoretical point of view, in practice it cannot be computed as already mentioned in Remark 3.29. Below, we provide particularly simple examples of classes of coherent programs, which follow mainly from Lemma 3.30. For instance,

Lemma 3.46 *Suppose that p is a conservative program. If, for every two coinitial transitions labeled respectively by A and B, we have $\mathrm{FV}(A) \cap \mathrm{FV}(B) = \emptyset$, then the program p is coherent.*

In our illustrative programming language, there is only one kind of variable and it contains integers, but it would be easy to extend it with another kind of variable which contains boolean values. We thus suppose, in the rest of this section that we can also store boolean values in variables: otherwise, the definition of the following procedure, which transforms a program into a coherent one, would be unnecessarily complicated.

Definition 3.47 Suppose given a conservative program p and a mutex a not occurring in p. We define a new program $\lceil p \rceil$ inductively by

$$
\begin{aligned}
\lceil A \rceil &= P_a; A; V_a & \lceil \texttt{skip} \rceil &= \texttt{skip} \\
\lceil P_b \rceil &= P_b & \lceil V_b \rceil &= V_b \\
\lceil p; q \rceil &= \lceil p \rceil; \lceil q \rceil & \lceil p \parallel q \rceil &= \lceil p \rceil \parallel \lceil q \rceil \\
\lceil \texttt{if } b \texttt{ then } p \texttt{ else } q \rceil &= \lceil \texttt{x:=}b \rceil; \texttt{if x then } \lceil p \rceil \texttt{ else } \lceil q \rceil \\
\lceil \texttt{while } b \texttt{ then } p \texttt{ else } q \rceil &= \lceil \texttt{x:=}b \rceil; \texttt{while x do } \lceil p \rceil
\end{aligned}
$$

Above, A denotes any action. In the cases for `if` and `while`, the variable x is supposed to be fresh, i.e., not used elsewhere.

Proposition 3.48 *Given a conservative program p, the program $\lceil p \rceil$ is coherent.*

The above transformation ensures that when an action is performed the mutex a is taken. This ensures that no two actions can be executed at the same time. Notice in particular that the transformation in the cases of `if` and `while` ensures that no other thread will be able to access the variables used by the boolean condition. Given a program p, the pruned asynchronous transition graph associated to $\lceil p \rceil$ has thus almost no independence tile, which explains why the above proposition is true: the only commutations are between the evaluation of the fresh variable corresponding to a conditional branching or a loop, and another action, but the freshness of the variable ensures coherence in this case.

Example 3.49 The program $p = \mathtt{x:=1} \parallel \mathtt{x:=2}$ is not coherent since the actions $\mathtt{x:=1}$ and $\mathtt{x:=2}$ do not commute: its associated pruned asynchronous transition graph is shown on the left below and the two distinct paths which are dihomotopic do not have the same semantics.

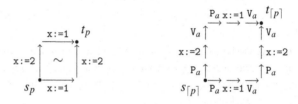

The transformed program $\lceil p \rceil$ is $(\mathrm{P}_a; \mathtt{x} := 1; \mathrm{V}_a) \parallel (\mathrm{P}_a; \mathtt{x} := 2; \mathrm{V}_a)$ and the associated pruned asynchronous transition graph is shown on the right: since there is no nontrivial pair of dihomotopic paths, coherence is immediate.

Example 3.50 Consider the program

$$p = \mathtt{x:=1} \parallel (\mathtt{if}\ \mathtt{x==0}\ \mathtt{then}\ \mathtt{y:=0}\ \mathtt{else}\ \mathtt{y:=1})$$

Its pruned asynchronous transition graph contains a tile of the form shown on the left and is therefore not coherent:

Namely, the path $\mathtt{x==0} . \mathtt{x:=1}$ is an execution trace because $\sigma_0(x) = 0$, whereas the path $\mathtt{x:=1} . \mathtt{x==0}$ is not an execution trace because the condition $\mathtt{x==0}$ is not satisfied. The transformed program $\lceil p \rceil$ is

```
(Pa;x:=1;Va) ||
(Pa;b:=(x==0);Va; if b then (Pa;y:=0; Va) else (Pa;y:=1; Va))
```

Even though the transformation removes most commutation tiles in the pruned asynchronous graph, it does not remove all of them, for instance it contains the subgraph shown on the right above.

Remark 3.51 The above translation seems to be particularly simple, but it is sometimes used in practice. This technique is for instance essentially the one used in the OCaml language because of the constraints imposed by the implementation of the garbage collector. Its documentation specifies [113]: "The OCaml run-time system is not reentrant: at any time, at most one thread can be executing OCaml code[...]. Technically, this is enforced by a *master lock* that any thread must hold while executing such code."

Remark 3.52 Of course this translation is by no means optimal. First, it adds lots of unnecessary blocking sections. For instance, if the program contains the instruction x:=1 and x is never used elsewhere, or the structure of the program is such that no other thread can access the variable x while it is assigned a value, then, following Lemma 3.30, it is unnecessary to enclose the action with P_a and V_a. Lemma 3.45 also shows that there is no need to add any blocking section if the program is sequential. The program transformation described in Definition 3.47 could also be made coherent in a more subtle way by using one mutex a_x for each variable x and enclosing every action A with P_{a_x} and V_{a_x} for each variable x occurring in A, such as in

$$P_{a_x}; P_{a_y}; P_{a_z}; x:=y+z; V_{a_x}; V_{a_y}; V_{a_z}$$

One can even go further by distinguishing, in every action x:=e, the variable x which is written to from the ones (occurring in e) that are just read; then removing a tile (of the pruned graph) whose edges are labeled by the actions x:=e and y:=e' only when $x \in FV(e)$ or $y \in FV(e')$. Doing so we would ensure that all dihomotopic execution traces are semantically equivalent, and thus avoid the unpleasant (undecidable) notion of coherent programs. Yet, we have explained earlier that, for sake of realism, we have decided to stick with the POSIX approach. It was motivated because most programming languages, contrarily to ours, have pointers: in this context, two distinct variables might refer to the same memory location, thus making the previous syntactic criterion irrelevant.

Example 3.53 (*Producer-consumer*) The coherence property for programs including more general data structures, such as queues, would require an extension of our toy language, as done in [51]. Nevertheless, for most practical applications, our current language is sufficient. The most well-known class of programs involving such data structures is the producer-consumer problem, which regulates the coordination between a number of distinguished processes p_i called *producers* (in practice, each producer emits some value v_i), and another number of distinguished processes c_i called *consumers* (which use the values emitted by producers in order, say, to update a global variable x). Each share a queue q, of capacity bounded by n: a *queue* is a data type which allows for atomically pushing up to n values (using the operation push_q) and retrieving them one by one (using pop_q) in the order they were put in. We suppose that our semantics was appropriately extended in order to support those. A classical way of ensuring the correct behavior of such a system is to use two resources e (for "not empty") and f (for "not full") of capacity n, see [33]: when a process acquires a lock on resource e (resp. f), it will be ensured that the queue is not empty (resp. not full). The resources are initialized so that f is not taken by any process, but e is considered to be with 0 as resource potential (no process can yet lock it unless some process releases it first). The producer and consumer processes are respectively

$$p_i = \mathrm{P}_f; \text{push}_q(v_i); \mathrm{V}_e \qquad\qquad c_i = \mathrm{P}_e; x := x \oplus \text{pop}_q(\,); \mathrm{V}_f$$

where $\oplus$ is an arbitrary associative and commutative operation. If we suppose that there are at least as many consumers than there are producers, the program is coherent (if we do not observe the contents of the queue). Below is depicted the asynchronous semantics of the program in a few cases, depending on the number of producers and consumers:

Notice that in the case with two producers and one consumer (on the right), the program is not coherent since x will contain either v_1 or v_2 in the terminal position, depending on the execution path taken to reach this position.

Remark 3.54 Usually, instead of having a fixed number of producers and consumers, there is one producer which iteratively produces values, i.e., its code is the same as above but encapsulated in a `while` loop, and similarly there is one consumer which iteratively consumes values. Notice that the corresponding program is not conservative, and hence cannot be described in the formalism of the book. Such a program could still be taken in account by rewriting the producer-consumer code using *monitors* [92], and extending the techniques developed here to handle those, but this is outside the scope of this book.

3.3.4 Programs with Mutexes Only

In this section, we will be interested in programs where all the resources are mutexes, i.e., have capacity 1. These are interesting because mutexes are the most widely used synchronization primitive, and one can show interesting properties in this particular case. We begin by observing that the asynchronous semantics $\check{G}_p$ of a conservative program p always satisfies the following property:

Definition 3.55 An asynchronous graph has *uniquely closing tiles* when given a pair of solid edges as in the figures below, there is at most one pair of dotted edges such that there is a tile relating both paths:

In an asynchronous graph satisfying the above property, one can define the residual of a path after another which intuitively corresponds to what "remains" of a path once the other has been taken.

Definition 3.56 Given two coinitial paths $s : x \twoheadrightarrow y$ and $t : x \twoheadrightarrow z$, the *residual* t/s of t after s is the path defined by induction on both paths by

$$t/\varepsilon = t \qquad \varepsilon/s = s \qquad (e_2 . t')/(e_1 . s') = ((t/e_1') . e_2')/s$$

where the last case is defined only if the transitions e_1 and e_2 can be closed as a tile of the form (3.4):

$$
\begin{array}{c}
t' \left| \begin{array}{cc} & t'/e_1' \\ e_1' & \\ e_2 & \sim \quad e_2' \\ & e_1 \quad s' \end{array} \right| ((t'/e_1') . e_2')/s'
\end{array}
\qquad \text{whenever} \qquad
\begin{array}{l}
s = e_1 . s' \\
t = e_2 . t'
\end{array}
$$

Lemma 3.57 *Given two coinitial paths s and t, the residual t/s is defined if and only if s/t is, and we have $s . (t/s) \sim t . (s/t)$.*

When the program p contains only mutexes, its asynchronous semantics can be shown to moreover satisfy the two following cube properties:

Definition 3.58 An asynchronous graph satisfies the *forward cube property* when whenever it contains an asynchronous subgraph as on the left, it also contains a subgraph as on the right (notice that the two "external" paths, $e_1 . e_2 . e_3$ and $e_3' . e_2' . e_1'$, are the same in both graphs):

$$
\begin{array}{ccc}
e_3 \nearrow \quad \searrow e_1' & & e_3 \nearrow \quad \searrow e_1' \\
e_2 \uparrow \sim \nearrow \uparrow e_2' & \Rightarrow & e_2 \uparrow \nwarrow \sim \nearrow \uparrow e_2' \\
e_1 \searrow \nearrow e_3' & & e_1 \searrow \quad \nearrow e_3' \\
& & \sim
\end{array}
$$

The right-to-left implication is called the *backward cube property*.

Proposition 3.59 *An asynchronous graph, whose tiles close uniquely and which satisfies the forward cube property also satisfies the following.*

1. *Given coinitial paths s, s', t, if $s \sim s'$ then $t/s = t/s'$ (both residuals are simultaneously defined or not).*
2. *Given coinitial paths s, t, t', if $t \sim t'$ then $t/s \sim t'/s$ (both residuals are simultaneously defined or not).*
3. *Given paths s, t, t', if $s.t \sim s.t'$ then $t \sim t'$.*
4. *Given paths s, s', t, t' such that $s.s' \sim t.t'$, the residuals s/t and t/s are both defined and there exists a path u such that $(t/s).u \sim s'$ and $(s/t).u \sim t'$*

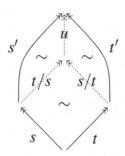

and moreover such a path u is unique up to dihomotopy.

Dual properties are satisfied when the backward cube property is verified.

Proof The proofs of 1, 2, and 3 are performed by induction on the derivation of dihomotopy and the length of paths, using the forward cube property. In 4, the required morphisms are obtained by starting from the homotopy between $s.s'$ and $t.t'$, and replacing all possible half cubes using the forward cube property. Uniqueness up to homotopy of u follows from 3.   □

Two coinitial morphisms f and g in a category are *compatible* when there exists morphisms f' and g' such that $f' \circ f = g' \circ g$ (and cocompatibility is defined dually). From the previous proposition, wet get, immediately, the following:

Corollary 3.60 *Given a conservative program p with mutexes only, the fundamental category $\vec{\Pi}_1(\check{G}_p)$ has pushout of compatible morphisms, pullbacks of cocompatible morphisms, and every morphism is epi and mono.*

Another interesting observation is that in this case, the homotopy classes of paths are generated by posets in the following sense. To every finite poset $(E, \leq)$ one can associate an asynchronous graph whose vertices are the downward closed subsets of E, called its *configurations*, there is an edge from x to y whenever $y = x \sqcup \{e\}$ for some $e \in E$, and all possible squares are filled with tiles. For instance, the poset whose Hasse diagram is shown on the left (elements are increasing from bottom to top) generates the asynchronous graph on the right:

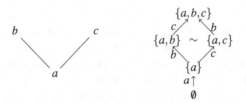

The generated asynchronous graph satisfies the forward and backward cube property, and tiles close uniquely. Moreover, all the paths from the initial configuration $\emptyset$ to the terminal configuration E are homotopic. Conversely, in an asynchronous graph satisfying those properties, every path homotopy class is generated by a partial order in this way: the elements of this poset are the *events*, i.e., the equivalence classes of transitions by the smallest equivalence relation identifying two transitions which occur as parallel sides of an independence tile such as e_1 and e_1' in (3.4).

In the above construction, the partial order expresses the dependencies between events; it can also be extended to the case where some events are incompatible, giving rise to the notion of *event structure* [165]. Links between event structures and other models such as asynchronous graphs are investigated in [126, 133, 167] and extended to precubical models in [73]. A similar (and strongly related) cube condition was introduced by Gromov in order to characterize *cubical complexes* (which are roughly geometric realizations of precubical sets) of non-positive curvature [78].

3.4 Cubical Semantics

3.4.1 Precubical Sets

In the previous section, we have enriched the structure of a graph in order to take commutation of two actions into account. We now generalize this idea to n actions. For instance, the program $\mathrm{P}_a; \mathrm{V}_a \parallel \mathrm{P}_a; \mathrm{V}_a \parallel \mathrm{P}_a; \mathrm{V}_a$ where a is a mutex such that $\kappa_a = 2$ generates the following pruned asynchronous transition graph:

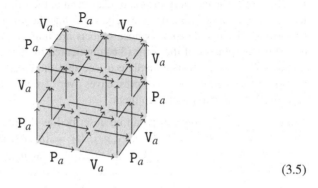

$$(3.5)$$

where all the squares are filled, but the interior of the cube is empty. Notice that there is no vertex in the middle and the figure can thus be seen as an empty subdivided cube. In the previous section, we saw the importance of distinguishing between empty and filled squares. We would like to extend here this methodology to all n-dimensional cubes: for instance, we would like to distinguish between an empty n-cube and a filled one. In order to formalize this, we use a generalization of asynchronous graphs called precubical sets. An asynchronous graph consists of three kinds of objects: 0-dimensional ones (the vertices), 1-dimensional ones (the edges), and 2-dimensional ones (the independence tiles). A precubical set will consists of sets of n-dimensional cubes for each $n \in \mathbb{N}$, together with their faces: each n-dimensional cube has $2n$ faces, i.e., a front and a back face in each direction i with $0 \le i < n$.

Definition 3.61 A **precubical set** C consists of a family $(C_n)_{n \in \mathbb{N}}$ of sets, whose elements are called n-cubes together with for all indices $n, i \in \mathbb{N}$ with $0 \le i < n$, maps

$$\partial_{n,i}^- : C_n \to C_{n-1} \quad \text{and} \quad \partial_{n,i}^+ : C_n \to C_{n-1} \tag{3.6}$$

respectively associating to an n-cube its *back* and *front face* in the ith direction, such that

$$\partial_{n,j}^\beta \partial_{n+1,i}^\alpha = \partial_{n,i}^\alpha \partial_{n+1,j+1}^\beta \tag{3.7}$$

for $0 \le i \le j < n$ and $\alpha, \beta \in \{-, +\}$. A *morphism* $f : C \to D$ between a precubical set C and a precubical set D consists of a family $(f_n : C_n \to D_n)_{n \in \mathbb{N}}$ of functions such that for every integers $n, i \in \mathbb{N}$ with $0 \le i < n$ and $\alpha \in \{-, +\}$,

$$\partial_{n,i}^\alpha \circ f_n = f_{n-1} \circ \partial_{n,i}^\alpha$$

The 0-cubes and 1-cubes of a precubical set are often called its *vertices* and *edges* respectively. Given a set $\mathscr{L}$ of labels, a *labeled* precubical set (C, ℓ) consists of a precubical set C together with a function $\ell : C_1 \to \mathscr{L}$ such that

$$\ell \circ \partial_{2,0}^- = \ell \circ \partial_{2,0}^+ \quad \text{and} \quad \ell \circ \partial_{2,1}^- = \ell \circ \partial_{2,1}^+$$

Remark 3.62 The category of precubical sets and their morphisms can be reformulated as a category of functors. Namely, we define the *precubical category* $\square$ as the opposite of the free category whose objects are integers and morphisms are generated by morphisms of the form (3.6) quotiented by the congruence generated by relations (3.7). It is then immediate to see that the category of precubical sets is isomorphic to the category $\hat{\square}$ of presheaves over $\square$, i.e., of functors $\square^{op} \to \mathbf{Set}$ and natural transformations between them.

Example 3.63 A Möbius strip can be described as the precubical set M such that $M_0 = \{x, x_1, x_2\}$, $M_1 = \{f_1, f_2, g, g_1, g_2\}$, $M_2 = \{h_1, h_2\}$ and $M_n = \emptyset$ for $n > 2$. The faces are given by $\partial_{1,0}^-(f_1) = x_1$, $\partial_{1,0}^+(f_1) = x$, $\partial_{1,0}^-(f_2) = x_2$, $\partial_{1,0}^+(f_2) = x$, $\partial_{1,0}^-(g_1) = x_1$, $\partial_{1,0}^+(g_1) = x_2$, $\partial_{1,0}^-(g) = x$, $\partial_{1,0}^+(g) = x$, $\partial_{1,0}^-(g_2) = x_2$, $\partial_{1,0}^+(g_2) = x_1$, $\partial_{2,0}^-(h_1) = f_1$, $\partial_{2,0}^+(h_1) = f_2$, $\partial_{2,1}^-(h_1) = g_1$, $\partial_{2,1}^+(h_1) = g$, etc. Graphically,

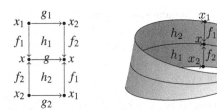

Notice that every graph $G = (V, \partial^-, \partial^+, E)$ can be seen as a precubical set C with $C_0 = V$, $C_1 = E$, and $C_n = \emptyset$ for $n > 1$, with maps $\partial^-_{1,0} = \partial^-$, $\partial^+_{1,0} = \partial^+$. An edge $e \in E$ thus has the following source and target:

$$\partial^-_{1,0}(e) \xrightarrow{\quad e \quad} \partial^+_{1,0}(e)$$

and this extends to labeled graphs and labeled precubical sets. Similarly, an asynchronous graph (G, I) can be seen as a precubical set C with C_0 and C_1 as above, $C_2 = I$ and $C_n = \emptyset$ for $n > 2$, with maps $\partial^-_{1,0}$ and $\partial^+_{1,0}$ as above, and given an element $h = (e_1 \cdot e'_2, e_2 \cdot e'_1)$ of I,

$$\partial^-_{2,0}(h) = e_1 \qquad \partial^+_{2,0}(h) = e'_1 \qquad \partial^-_{2,1}(h) = e_2 \qquad \partial^+_{2,1}(h) = e'_2 \qquad (3.8)$$

which corresponds to the following figure

$$\partial^-_{2,0}(h) \overset{\partial^+_{2,1}(h)}{\underset{\partial^-_{2,1}(h)}{\uparrow \quad h \quad \uparrow}} \partial^+_{2,0}(h)$$

It is easy to see that this provides a full and faithful embedding of the category of (asynchronous) graphs into the category of precubical sets: the morphisms between asynchronous graphs are in bijection with the morphisms between the corresponding precubical sets.

Example 3.64 The graph $I = x \bullet \xrightarrow{\quad f \quad} \bullet y$ with two vertices x and y, and one edge f from x to y, can be seen as the precubical set C with $C_0 = \{x, y\}$, $C_1 = \{f\}$ and $C_n = \emptyset$ for $n > 1$, and face maps given by $\partial^-_{1,0}(f) = x$ and $\partial^+_{1,0}(f) = y$.

Conversely, any labeled precubical set (C, ℓ) has an underlying asynchronous transition graph (G, I), with C_0 as vertices of the graph, C_1 as edges of the graph, with source and target respectively given by $\partial^-_{1,0}$ and $\partial^+_{1,0}$ and labels by ℓ, and there is a tile relating paths $e_1 \cdot e'_2$ and $e_2 \cdot e'_1$, as in (3.4), whenever there is a 2-cube $h \in C_2$ satisfying the relations (3.8). We can thus easily import concepts from Sect. 3.3.2. For instance,

Definition 3.65 A *path* in a precubical set is a finite sequence $e_1, \ldots, e_n$ of 1-cubes such that $\partial^+_{1,0}(e_i) = \partial^-_{1,0}(e_{i+1})$ for every index i with $1 \leq i < n$. The **dihomotopy** relation $\sim$ on paths is the smallest equivalence relation, which is a congruence w.r.t. concatenation, and such that $(e_1 \cdot e'_2) \sim (e_2 \cdot e'_1)$ whenever there is a 2-cube $h \in C_2$ satisfying the relations (3.8).

Similarly, the fundamental category of a precubical set C can be defined as the fundamental category of its underlying asynchronous graph:

Definition 3.66 The *fundamental category* $\vec{\Pi}_1(C)$ associated to a precubical set C is the category whose objects are the vertices of C and morphisms from x to y are paths from x to y up to dihomotopy.

The operations previously defined on asynchronous graphs can be extended without difficulty to precubical sets, while coinciding with previous operations on the embedding of asynchronous graphs. Given two precubical sets C and D,

- their *disjoint union* is the precubical set $C \sqcup D$ defined by

$$(C \sqcup D)_n = C_n \sqcup D_n$$

  with boundary maps induced by those of C and D,
- their *tensor product* is the precubical set $C \otimes D$ defined by

$$(C \otimes D)_n = \coprod_{i+j=n} C_i \times D_j$$

  with boundary map $\partial^\alpha_{n,k} : (C \otimes D)_n \to (C \otimes D)_{n-1}$ defined on $(x, y) \in C_i \times D_j$, with $i + j = n$, by

$$\partial^\alpha_{n,k}(x, y) = \begin{cases} (\partial^\alpha_{n,k}(x), y) & \text{if } 0 \leq k < i \\ (x, \partial^\alpha_{n,k-i}(y)) & \text{if } i \leq k < n \end{cases}$$

and other operations such as quotient or restriction can be defined similarly (when restricting to a subset of vertices, one has to remove all cubes which admit, as iterated face, a vertex in the complement of this subset). Moreover, these operations extend as expected to labeled precubical sets.

Example 3.67 We write I for the precubical set described in Example 3.64, corresponding to the graph with one edge $f : x \to y$, and S^1 for the precubical set corresponding to the graph with one vertex z and one edge g (from z to z):

(notice that we could also have defined S^1 as $I[x = y]$). The precubical set C corresponding to an empty cylinder can be obtained as $C = I \otimes S^1$. Namely, we have $C_0 = \{(x, z), (y, z)\}$, $C_1 = \{(f, z), (x, g), (y, g)\}$, $C_2 = \{(f, g)\}$ and $C_n = \emptyset$ for $n > 2$. Faces are given by $\partial^-_{1,0}(f, z) = (x, z)$, $\partial^+_{1,0}(f, z) = (y, z)$, $\partial^\alpha_{1,0}(x, g) = (x, z)$, $\partial^\alpha_{1,0}(y, g) = (y, z)$, $\partial^-_{2,0}(f, g) = (x, g)$, $\partial^+_{2,0}(f, g) = (y, g)$ and $\partial^\alpha_{2,1}(f, g) = (f, z)$, with $\alpha \in \{-, +\}$. Graphically:

$$C \quad = \quad \begin{matrix} & (x,z) \ (f,z) \ (y,z) \\ (x,g) & (f,g) & (y,g) \end{matrix}$$

Many others classical examples arise from simple precubical sets using tensor products. For instance, the square is $I \otimes I$, more generally an n-cube is $I^{\otimes n}$ (the tensor product of n copies I), an empty torus is $S^1 \otimes S^1$, a filled square toroid as $I \otimes I \otimes S^1$, etc.

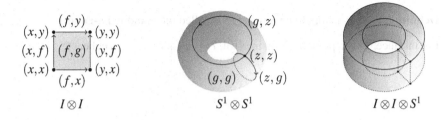

$$I \otimes I \qquad\qquad S^1 \otimes S^1 \qquad\qquad I \otimes I \otimes S^1$$

Using the above constructions, we can define models of concurrent programs in labeled precubical sets, generalizing the definitions given in Sect. 3.3.2

Definition 3.68 The *precubical transition set* C_p associated to a program p is then defined as in Definition 2.9, using the above operations on precubical sets which are labeled in $\mathscr{L} = \mathscr{C}_{\text{act}} \sqcup \mathscr{B} \sqcup \mathrm{P}_{\mathscr{R}} \sqcup \mathrm{V}_{\mathscr{R}}$. The *pruned precubical transition set* $\check{C}_p$, which will often be called the **cubical semantics** of p, is then obtained by restricting to valid vertices (i.e., 0-cubes), as in Definition 3.16

Example 3.69 Consider the following program:

$$p = \mathrm{P}_a; \mathrm{V}_a \parallel \mathrm{P}_a; \mathrm{V}_a \parallel \mathrm{P}_a; \mathrm{V}_a$$

Depending on the capacity of the resource a, the cubical semantics $\check{C}_p$ of p is as follows:

- if $\kappa_a = 3$ then $\check{C}_p$ is a subdivided filled cube,
- if $\kappa_a = 2$ then $\check{C}_p$ is a subdivided hollow cube,
- if $\kappa_a = 1$ then $\check{C}_p$ is a subdivided skeletal cube reduced to its edges,
- if $\kappa_a = 0$ then $\check{C}_p$ only consists of 8 vertices.

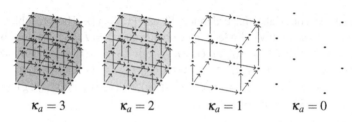

$$\kappa_a = 3 \qquad \kappa_a = 2 \qquad \kappa_a = 1 \qquad \kappa_a = 0$$

3.4.2 The Geometric Realization

In order to relate the algebraic approach developed in this chapter and the geometric models that will be presented in following chapters, we recall how a precubical set can be seen as a topological space obtained by gluing cubes according to the precubical set.

Definition 3.70 The topological space I^n is called the **standard n-cube**.

For instance, I^0 is a point, I^1 is an interval, I^2 is a square, I^3 is a cube, etc.

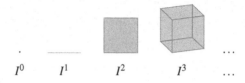

$$I^0 \qquad I^1 \qquad I^2 \qquad I^3 \qquad \ldots$$

Notice that I^n has $2n$ faces, a back and a front face in each direction $0 \le i < n$, which are $(n-1)$-cubes included in I^n that can be described as follows: given $0 \le i < n$, the ith back-face is given by the inclusion $\iota_{n,i}^- : I^{n-1} \to I^n$ defined by

$$\iota_{n,i}^-(x_0, \ldots, x_{n-1}) = (x_0, \ldots, x_{i-1}, 0, x_i, \ldots, x_{n-1})$$

and the ith front face is given by the inclusion $\iota_{n,i}^+ : I^{n-1} \to I^n$ defined similarly with 1 instead of 0. It is natural to think about the standard n-cube as a topological counterpart of an n-cube, and a precubical set as a gluing of such cubes:

Definition 3.71 The **geometric realization** of a precubical set C is the topological space

$$|C| = \coprod_{n \in \mathbb{N}} (C_n \times I^n) \, / \approx$$

where C_n is equipped with the discrete topology, and $\approx$ is the equivalence relation generated by relations $(\partial_{n,i}^\alpha(x), p) \approx (x, i_{n-1,i}^\alpha(p))$ for $n \in \mathbb{N}$, $x \in C_n$ and $p \in I^{n-1}$.

Example 3.72 The topological realization of the Möbius strip precubical set given in Example 3.63 is a Möbius strip (in the usual topological sense), and similarly for the other examples given in Example 3.67.

Remark 3.73 A more abstract point of view can be developed as follows. Recall from Remark 3.62 that precubical sets are presheaves over the category $\square$. Now, one can define a functor $I : \square \to \mathbf{Top}$, which sends an object n to I^n and generators $\partial_{n,i}^\alpha : (n-1) \to n$ of face maps to face maps $\iota_{n,i}^\alpha : I^{n-1} \to I^n$. The geometric realization can then be obtained as the colimit

$$|C| = \mathrm{colim}\left(y/C \xrightarrow{\pi} \square \xrightarrow{I} \mathscr{C}\right) \tag{3.9}$$

where $y : \square \to \hat{\square}$ is the Yoneda embedding, y/C is the slice category of the functor y over $C \in \hat{\square}$ (its objects are pairs (n,f) with $n \in \square$ and $f : yn \to C$, and it is sometimes called the category of elements of C), $\pi : y/C \to \square$ is the first projection functor (which is defined for any slice category) and $\mathscr{C} = \mathbf{Top}$. The interest of this abstract version is that it can be easily generalized, by replacing $\mathbf{Top}$ by any cocomplete category $\mathscr{C}$, and I by any functor $I : \square \to \mathscr{C}$: the definition (3.9) still makes sense and satisfies the same main properties as the usual geometric realization, as presented in Definition 3.71. This will be useful in order to study other models such as directed topological spaces (see Definition 4.10).

Unfortunately, the properties of this fundamental construction cannot be detailed here [56, 111]. We only mention that the geometric realization functor admits a right adjoint:

Proposition 3.74 *The geometric realization functor* $|-| : \hat{\square} \to \mathbf{Top}$ *admits a right adjoint, called the* nerve *functor* $N : \mathbf{Top} \to \hat{\square}$, *which is defined on objects* $X \in \mathbf{Top}$ *and* $n \in \square^{op}$ *by* $NXn = \mathbf{Top}(In, X)$.

Also, given a precubical set C, to every 0-cube $x \in C_0$ is canonically associated a point in $|C|$, that we denote by $|x|$, and similarly elements of C_1 can be seen as segments, etc.

3.5 Historical Notes

One of the most classical semantics for concurrency is based on automata, or transition systems [99, 122, 128, 143], and is known as *interleaving* semantics [5, 135], used heavily in the early days of process algebras (CCS, π-calculus, etc.) [93, 124]. Another classical semantics for concurrency originated in the work of Petri [139], in reaction to the predominant automata-theoretic approach to semantics, starting with the idea that automata "are not capable of representing the actual physical flow of information" [139] in concurrent and distributed systems.

From there on, a huge literature started on so-called *truly concurrent* semantics, where (sequential) non-determinism is considered as essentially different from the apparent non-determinism that a *sequential* observer [140] would see when two or more processes are executed concurrently. This has a number of advantages. First, most truly concurrent semantics for concurrency give rise to smaller models than interleaving semantics. It generally helps prevent the "state space explosion problem." Instead of representing the potentially exponential number of interleavings of independent actions, truly concurrent semantics generally encodes this implicitly as some form of independence relation, as in Mazurkiewicz trace theory [30, 121] or in asynchronous transition systems [70]. For instance, applications to efficient model-checking based on partial-order reduction [60] and on Petri nets [159] have been developed. Starting from the early 1990s, a series of seminal papers have been advocating to generalize those models by using precubical sets in order to study concurrency. Those have explored the use of precubical sets and *Higher-Dimensional Automata* (which are labeled precubical sets equipped with a distinguished beginning vertex) [141, 160], have begun to suggest possible homology theories [63, 72], and have found applications to serializability [82]. Notice that (pre)cubical sets, which are natural in our context, are also classical objects in algebraic topology (although somewhat less classical than simplicial sets), from the early work of Kan in algebraic topology starting from the 1940s and the thesis of Serre [153], to the Bangor group [20], the work of Jardine [95], to important aspects of the proof of the Poincaré conjecture (special cube complexes, see for instance [83]). There are also links with other fields of computer science that will not be developed in this book, and are only evoked in the concluding Chap. 8. Let us mention rewriting theory, links with Squier's theorem [3, 157], models for higher-order type theory [10], bisimulation semantics [39], etc.

A more domain theoretic approach, as opposed to the operational semantics approach as in the above, has been introduced in [132], with links to Petri Nets, under the name of (prime) event structures. A good reference for the relationships between all these models, and in particular asynchronous graphs, can be found in [167], and has been generalized to precubical models in [73]. These representations are generally also more modular. It has been advocated for instance [162] that these semantics may help in deriving properties of concurrent systems at different levels of abstractions (and not by having to enumerate all potential actions, as in interleaving semantics). Second, it allows for describing finer properties. Obviously, these semantics distinguish concurrent executions from mutual exclusions, and for some, they even distinguish all kinds of weaker synchronizations (e.g., counting semaphores as we will see later in the book) and the number of concurrent processes which are scheduled at the same time on distinct processors. In short, general truly concurrent semantics generalize properties that can be observed, from sequential to concurrent observers. This is instrumental in a number of applications, such as in fault-tolerant protocols for distributed systems, which will not be handled here, see [88] and Chap. 8.

Chapter 4
Directed Topological Models of Concurrency

In this chapter, we continue the presentation of various models for concurrent programs and, in particular, study topological models, with the aim of importing tools and techniques coming from algebraic topology in order to ease verification of concurrent programs. In those models, the state space of a program is described as a topological space, and an execution naturally corresponds to a path in this space. However, usual topological spaces are not completely suitable for our purposes, because they do not take the causality of the program into account: the execution of a program can only go forward in time (a program cannot execute some actions backward), whereas there is no corresponding constraint on the paths in a topological space. In order for the models to behave properly, we are led to enrich the concept of a topological space so that it takes causality into account. We shall then focus our attention on *directed* paths, i.e., the ones respecting causality. Many variants of this notion have been proposed, but we will mainly focus here on *d-spaces* as introduced by Grandis, because they are technically more tractable and a widely accepted notion nowadays. We only provide a brief introduction to those here (Sect. 4.1), the reader interested in more detail is advised to consult the reference book about the subject [77]. In particular, we explain how various classical notions in algebraic topology extend to this setting: homotopy and the fundamental category (Sect. 4.2), the Seifert–van Kampen theorem (Sect. 4.3.1), universal covering spaces (Sect. 4.3.2), and hint at various other constructions (Sect. 4.4).

4.1 Directed Spaces

We first recall that the *unit interval* is the topological space $I = [0, 1]$ equipped with the Euclidean topology (open sets are generated by open intervals). A *path* f in a topological space X is a continuous map $f : I \to X$, the point $f(0)$ is the *source* and $f(1)$ is the *target* of the path, and we sometimes write $f : f(0) \to f(1)$ to indicate those endpoints. A *loop* is a path with the same source and target. The constant path, whose image is a given point x, is written ε_x. Given two paths $f, g : I \to X$, their

© Springer International Publishing Switzerland 2016
L. Fajstrup et al., *Directed Algebraic Topology and Concurrency*,
DOI 10.1007/978-3-319-15398-8_4

concatenation is the path $f \cdot g$ such that $f \cdot g(t)$ is $f(2t)$ if $0 \leqslant t \leqslant 1/2$ and is $g(2t - 1)$ otherwise. A path $f : I \to I$ is a directed *partial reparametrization* if it is weakly increasing (and a *reparametrization* whenever $f(0) = 0$ and $f(1) = 1$).

4.1.1 A Definition

A directed topological space is a topological space equipped with a coherent set of paths that are considered as directed:

Definition 4.1 A **d-space** (X, dX) consists of a topological space X together with a set dX of paths of X, the *directed paths* or *d-paths* or *dipaths*, such that

1. every constant path is directed,
2. the precomposition of a directed path with a directed partial reparametrization is directed,
3. the concatenation of two directed paths is directed.

A morphism of d-spaces (or *d-map*) $h : X \to Y$, is a continuous map $h : X \to Y$ which preserves directed paths, in the sense that for every directed path $f \in dX$ we have $h \circ f \in dY$. We write **dTop** for the category of d-spaces and d-maps.

Since partial reparametrizations are not necessarily surjective, the second condition implies in particular that dX is closed under taking subpaths. A *subspace* (Y, dY) of a d-space (X, dX) is a subset Y of X that inherits its topology and d-paths from X, i.e., $dY = \{f \in dX \mid f(I) \subseteq Y\}$ (in fact, every subset of a d-space can be equipped with a d-space structure in this way).

Example 4.2 The fundamental example of a d-space is the *directed unit interval* $\vec{I} = (I, dI)$ where $I = [0, 1]$ is the unit interval and dI is the set of paths $f : I \to I$ which are weakly increasing, i.e., the partial reparametrizations. Notice that the set dX of directed paths of a directed space X is in bijection with d-maps from $\vec{I}$ to X, i.e., $dX \cong \textbf{dTop}(\vec{I}, X)$. The d-space structure on a topological space is of course not unique. For instance, the set of constant paths in I or the set of all paths in I also define a d-space structure on I, see Proposition 4.6.

Example 4.3 A topological space X equipped with a partial order, whose graph is a closed subspace of $X \times X$, is called a **pospace**. The condition imposes that the partial order is compatible with the topology of the space: for instance, it ensures that the limit of an increasing sequence of points is above all the points in the sequence, or that limit is compatible with pointwise ordering of sequences. Every pospace defines a d-space by defining dX as the set of weakly increasing paths $f : I \to X$. The directed unit interval $\vec{I}$ of Example 4.2 is an instance of this construction, starting from the pospace $I = [0, 1]$ equipped with the usual partial order. Similarly, we write $\mathbb{R}$ for the *directed real line*. The *directed standard n-cube* $\vec{I}^n$ is the d-space generated by I^n equipped with the product order (i.e., $(x_1, \ldots, x_n) \leqslant (y_1, \ldots, y_n)$ iff for each index i, we have $x_i \leqslant y_i$).

Example 4.4 The *directed complex plane* $\vec{\mathbb{C}}$ is the complex plane $\mathbb{C}$ equipped with the set $d\mathbb{C}$ of directed paths consisting of paths $f : I \to \mathbb{C}$ such that the function $t \mapsto |p(t)|$ is (weakly) increasing and $t \mapsto \arg(p(t))$ is (weakly) increasing modulo 2π: increasing paths are those which are turning counterclockwise and going further away from the origin. The *directed circle* $\vec{S}^1$ is the subspace consisting of points $z \in \mathbb{C}$ such that $|z| = 1$, as figured on the left:

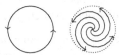

Similarly, the *directed disk* $\vec{D}^2$ is the subspace of $\vec{\mathbb{C}}$ of points z satisfying $|z| \leqslant 1$ and is pictured on the right. Notice that neither of these spaces are generated from a topological space equipped with a partial order as described in Example 4.3. One can further remark that every point of the directed circle admits a neighborhood homeomorphic to $\vec{I}$, on which the direction arises from a partial order, but this is not the case for the directed complex plane: there is no neighborhood of the origin whose direction is generated by a partial order (we have illustrated some directed paths starting from the origin on the right). Such a point (with the property that every neighborhood contains a nontrivial loop) is called a *vortex*.

Further examples of d-spaces can be constructed from the above ones using limits and colimits described in next section.

4.1.2   Limits and Colimits

A major property of the category **dTop** is that it has all limits and colimits (we refer the reader to Grandis' book [77] for details and proofs):

Proposition 4.5 *The category* **dTop** *is both complete and cocomplete.*

In the nondirected setting, i.e., for the category **Top** of topological spaces and continuous maps, this is also true and well studied: roughly, colimits correspond to gluing spaces, whereas limits correspond to taking some products of spaces. A first intuition about what (co)limits look like in the directed setting is given by the fact that they coincide with usual (co)limits on the underlying topological spaces:

Proposition 4.6 *The forgetful functor* $U : $ **dTop** $\to$ **Top**, *which is defined on objects by* $U(X, dX) = X$, *has both a left and a right adjoint. It thus preserves colimits and limits.*

Proof Given a topological space X, the collection of all constant paths (resp. all paths) on X equips it with a d-space structure, and this operation extends to a functor which is left (resp. right) adjoint to the forgetful functor. $\square$

Note that the previous proposition allows us to equip any topological space with two possible canonical d-space structures. In particular, in the following, the *unit interval* I will implicitly be seen as equipped with all paths as directed paths, except when it occurs at the left of an arrow in which case only constant paths are directed: in the d-space $I \times I$ every path is directed, but a map $f : I \to X$ denotes a nondirected path in a d-space X.

Some usual (co)limits, which are topological analogs of the operations on graphs provided in Sect. 2.2.1, can be described explicitly as follows:

- The *terminal d-space* 1 is the d-space containing only one point.
- The *cartesian product* $X \times Y$ of two d-spaces X and Y has $d(X \times Y) = dX \times dY$.
- The *disjoint union* $X \sqcup Y$ of two d-spaces X and Y is such that $d(X \sqcup Y) = dX \sqcup dY$.
- The *quotient* $X[x = y]$ of a d-space X by identifying two points x and y is the d-space X where x and y have been identified, and a directed path is, up to reparametrization, a finite sequence of directed paths $(f_i : z_i \twoheadrightarrow z_i')_{1 \leqslant i \leqslant n}$ in X such that $z_i, z_i' \in \{x, y\}$, except possibly z_1 and z_n', with $z_i' \neq z_{i+1}$.

Example 4.7 The product $\vec{S}^1 \times \vec{I}$ is the directed empty cylinder, drawn on the left. Similarly, the product $\vec{D}^2 \times \vec{I}$ is the directed filled cylinder. The product $\vec{S}^1 \times \vec{S}^1$ is the empty torus shown on the right.

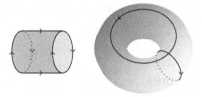

Example 4.8 Colimits in the category **dTop** do not always give the "expected" result in the presence of vortexes. For instance, consider the pushout of the diagram

$$\vec{S}^1 \times \vec{I} \longleftarrow \vec{S}^1 \longrightarrow 1$$

where the arrow pointing to the left is the inclusion of the directed circle at the base of the directed empty cylinder (see Example 4.7), and the arrow pointing to the right is the terminal arrow: the resulting space is obtained by squashing the base of a cylinder to a point. We could expect the result to be the directed disk $\vec{D}^2$ (see Example 4.4), but this is not the case: while the underlying topological space is the same, an "upward spiral" (such as the path $t \mapsto t e^{i/t}$, whose image at 0 is 0) cannot be written as a finite concatenation of dipaths winding only once around the origin, it is therefore not directed whereas it belongs to $d\vec{D}^2$. For a workaround, see [87].

Given a directed topological space Y, the set $Y^{\vec{I}}$ of d-maps from $\vec{I}$ to Y is isomorphic to the set dY as noted in Example 4.2. It can be equipped with the *compact-open topology*, which is the topology generated by the sets of functions $f : \vec{I} \to Y$ such that $f(K) \subseteq U$ for some compact $K \subseteq I$ and open $U \subseteq Y$ (these form a subbasis for the topology). From the resulting topological space, we can define a d-space by declaring that the directed paths $h : I \to Y^{\vec{I}}$ in $dY^{\vec{I}}$ are the continuous functions such that the function $t \mapsto h(t)(u)$ is a directed path in Y for every $u \in \vec{I}$. We will always implicitly equip $Y^{\vec{I}}$ with this d-space structure, which is justified by the following property, showing that $\vec{I}$ is *exponentiable*:

Proposition 4.9 *For all d-spaces X and Y, there is a natural bijection between* $\mathbf{dTop}(X \times \vec{I}, Y)$ *and* $\mathbf{dTop}(X, Y^{\vec{I}})$.

Two directed paths f and g in a d-space X can be seen as points in $X^{\vec{I}}$.

The standard directed n-cube $\vec{I}^n$ defined in Example 4.3 is the space obtained as the product of n copies of $\vec{I}$. As in Sect. 3.4, we can define a functor $\vec{I} : \square \to \mathbf{dTop}$ which to every object n associates $\vec{I}^n$ and images of morphisms are defined as in the nondirected case. Because the category $\mathbf{dTop}$ is cocomplete, this induces a functor $|-| : \hat{\square} \to \mathbf{dTop}$ associating a directed topological space to every precubical set, see Remark 3.73.

Definition 4.10 The **directed geometric realization** functor $|-| : \hat{\square} \to \mathbf{dTop}$ is the functor defined on a precubical set C by

$$|C| = \coprod_{n \in \mathbb{N}} (C_n \times \vec{I}^n)/\approx$$

where the equivalence relation $\approx$ is defined as in Definition 3.71. In the above formula, C_n is equipped with the discrete topology and the discrete d-space structure (i.e., only constant paths are directed).

Example 4.11 The geometric realization of the precubical set corresponding to the cylinder and the torus given in Example 3.67 are the directed torus cylinder and torus described in Example 4.7.

Proposition 4.12 *The directed geometric realization functor preserves all colimits and sends tensor products of finite precubical sets to cartesian products.*

Proof By Proposition 3.74, the geometric realization functor admits a right adjoint and thus preserves colimits by the Freyd adjoint functor theorem [120]. The proof of the second part of the proposition is more involved and we refer to [56] for a proof. □

4.1.3 Directed Geometric Semantics

The constructions given in the previous section provide us with tools to define a semantics for programs in directed topological spaces, mimicking the definition provided for graphs in Sect. 2.2.2 (see Definition 2.9, in particular for illustrations of the different cases).

Definition 4.13 Suppose an operational semantics is given. To any conservative program p, we associate a quadruple (G_p, s_p, t_p, r_p) consisting of a directed topological space G_p together with two points $s_p, t_p \in G_p$, the *beginning* and the *end*, and a function $r_p : G_p \to (\mathscr{R} \to \mathbb{Z})$, the *resource potential*, defined inductively as follows:

- action: given $A \in \mathscr{C}_{\mathrm{act}}$,

$$G_A = \vec{I} \qquad s_A = 0 \qquad t_A = 1 \qquad r_A(x)(a) = 0$$

- locking: given $a \in \mathscr{R}$,

$$G_{\mathrm{P}_a} = \vec{I} \qquad s_{\mathrm{P}_a} = 0 \qquad t_{\mathrm{P}_a} = 1 \qquad r_{\mathrm{P}_a}(x)(b) = \begin{cases} 0 & \text{if } b \neq a \text{ or } x \leqslant 0.5 \\ -1 & \text{otherwise} \end{cases}$$

- unlocking: given $a \in \mathscr{R}$,

$$G_{\mathrm{V}_a} = \vec{I} \qquad s_{\mathrm{P}_a} = 0 \qquad t_{\mathrm{P}_a} = 1 \qquad r_{\mathrm{V}_a}(x)(b) = \begin{cases} 0 & \text{if } b \neq a \text{ or } x < 0.5 \\ 1 & \text{otherwise} \end{cases}$$

- skip: writing $1 = \{*\}$ for the terminal d-space,

$$G_{\mathrm{skip}} = 1 \qquad s_{\mathrm{skip}} = * \qquad t_{\mathrm{skip}} = * \qquad r_{\mathrm{skip}}(a)(x) = 0$$

- sequence:

$$G_{p;q} = (G_p \sqcup G_q)\,[t_p = s_q] \qquad s_{p;q} = s_p \qquad t_{p;q} = t_q$$

$$r_{p;q}(x)(a) = \begin{cases} r_p(x)(a) & \text{if } x \in G_p \\ r_q(x)(a) + r_p(t_p)(a) & \text{if } x \in G_q \end{cases}$$

- conditional branching: with $p = \texttt{if } b \texttt{ then } p_1 \texttt{ else } p_2$,

$$G_p = (G_b \sqcup G_{\neg b} \sqcup G_{p_1} \sqcup G_{p_2})[s_b = s_{\neg b}, t_b = s_{p_1}, t_{\neg b} = s_{p_2}, t_{p_1} = t_{p_2}]$$

$$s_p = s_b \qquad t_p = t_{p_1} \qquad r_p(x)(a) = \begin{cases} 0 & \text{if } x \in G_b \text{ or } x \in G_{\neg b} \\ r_{p_1}(x)(a) & \text{if } x \in G_{p_1} \\ r_{p_2}(x)(a) & \text{if } x \in G_{p_2} \end{cases}$$

- conditional loop: with $p = \texttt{while } b \texttt{ do } q$,

$$G_p = (G_b \sqcup G_{\neg b} \sqcup G_q)\,[t_b = s_q, s_b = t_q, s_{\neg b} = t_q] \qquad s_p = s_b \qquad t_p = t_{\neg b}$$

$$r_p(x)(a) = \begin{cases} 0 & \text{if } x \in G_b \text{ or } x \in G_{\neg b} \\ r_q(x)(a) & \text{if } x \in G_q \end{cases}$$

- parallel:

$$G_{p\|q} = G_p \times G_q \qquad s_{p\|q} = (s_p, s_q) \qquad t_{p\|q} = (t_p, t_q)$$
$$r_{p\|q}(x, y)(a) = r_p(x)(a) + r_q(y)(a)$$

In the above, given a condition b, G_b denotes $\vec{I}$ with s_b (resp. t_b) as source (resp. target). The *forbidden region* is the subspace

$$R_p = \{x \in G_p \mid \exists a \in \mathcal{R}, r_p(x)(a) + \kappa_a < 0 \text{ or } r_p(x)(a) > 0\}$$

The **geometric semantics** $\check{G}_p$ of a program p is the d-space defined as the subspace $\check{G}_p = G_p \backslash R_p$ of G_p.

Remark 4.14 The assumption that the program p is conservative is crucial in order to show that the definition of the resource function makes sense.

The geometric semantics can be seen as a "continuous" version of the semantics developed up to now. It associates to each program a d-space whose directed paths correspond to executions of programs. The possibility of continuously deforming a path into another will mean that they are equivalent in the sense that they only differ by exchanging commuting actions, see Sect. 3.3.1. A simple example is the geometric semantics of a program of the form $A \, ; B$, as shown on the left:

The space $\check{G}_{A\,;B}$ is the directed interval $[0, 2]$, with the point 0 as beginning and 2 as end. A directed path $f : \vec{I} \to [0, 2]$ in this space should be thought as an execution of the program: the point $p(t) \in \check{G}_{A\,;B}$ will go increasingly from 0 to 2 when its argument t (corresponding to the current execution time) goes from 0 to 1. The action A (resp. B) is performed precisely when $p(t)$ is equal to 0.5 (resp. 1.5), which is why we distinguished and labeled these points in the figure. For a slightly more interesting example, consider the geometric semantics of the program $p = (A \, ; B) \parallel C$, shown on the right (ignore the two dotted and dashed paths for now): we have $\check{G}_{(A\,;B)} \parallel C = \check{G}_{A\,;B} \times \check{G}_C = [0, 2] \times [0, 1]$, the beginning being $(0, 0)$ and the end $(2, 1)$. A directed path $f : \vec{I} \to \check{G}_p$ is a pair $f(t) = (f_1(t), f_2(t))$, parametrized by a time $t \in \vec{I}$, of points in $f_1(t) \in \check{G}_{A\,;B}$ and $f_2(t) \in \check{G}_C$ which are both increasing as time increases, and the action A (resp. B, resp. C) is performed when $f_1(t) = 0.5$ (resp. $f_1(t) = 1.5$, resp. $f_2(t) = 0.5$). We have drawn two such paths in the figure: in the dotted one, A is performed, followed by B and then C (i.e., it corresponds to the execution trace $A.B.C$) and the dashed one corresponds to the execution trace $C.A.B$. Note that there are also such paths for which, at some instant t, $f_1(t) = f_2(t) = 0.5$, i.e., the actions A and C are performed *at the same time*. One should not give too much importance to the choice for the coordinates in the semantics (e.g. 0.5 for A) or more generally the length. We could have as well adopted other conventions, resulting in isomorphic spaces: what really matters here is the relative position of the actions. Thus, in the following, we sometimes assimilate actions with their coordinates and for instance write (A, C) for the point $(0.5, 0.5)$ in the above example.

Example 4.15 Consider the program $p = \mathsf{P}_a ; \mathsf{V}_a$. We write J for the pospace $[0, 2]$ and R for the pospace $]\frac{1}{2}, \frac{3}{2}[$. With $\kappa_a = 1$, the geometric semantics of the programs $p \parallel p$ and $p \parallel p \parallel p$ are, respectively,

$$
\begin{aligned}
\check{G}_{p\parallel p} &= (J \times J)\backslash(R \times R) \\
\check{G}_{p\parallel p\parallel p} &= (J \times J \times J)\backslash((J \times R \times R) \cup (R \times J \times R) \cup (R \times R \times J))
\end{aligned}
$$

With $\kappa_a = 2$ the situation is quite different since the geometric semantics of $p \parallel p$ is $J \times J$ and the one of $p \parallel p \parallel p$ is $(J \times J \times J)\backslash(R \times R \times R)$. These are represented in the table below:

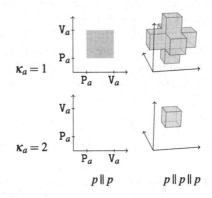

$$p \parallel p \qquad\qquad p \parallel p \parallel p$$

For obvious reasons, the program on the bottom right is often called the "floating cube".

As shown in the above example, we often draw the forbidden area using a grayed region.

Example 4.16 Consider the program $p = \mathsf{P}_a ; (\text{if } b \text{ then } \mathsf{V}_a \text{ else } \mathsf{V}_a)$, where a is a mutex. Its geometric semantics is the graph

$$
\check{G}_p \quad = \quad s_p \xrightarrow{\ \ \mathsf{P}_a\ \ } \begin{array}{c} \overset{b}{\diagup} \overset{\mathsf{V}_a}{\diagdown} \\ \underset{\neg b}{\diagdown} \underset{\mathsf{V}_a}{\diagup} \end{array} \xrightarrow{\ \ } t_p
$$

Note that the interior of the square above is not filled: there are two maximal dipaths.

Example 4.17 The geometric semantics of the programs $p = A; \text{while } b \text{ do } B$, and $q = (\mathsf{P}_a; \text{ while } b \text{ do } (\mathsf{V}_a; \mathsf{P}_a)) \parallel (\mathsf{P}_a; \mathsf{V}_a)$, where a is a mutex, are, respectively,

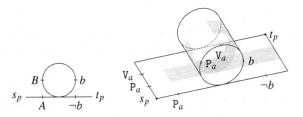

Example 4.18 (*Dining philosophers*) The geometric semantics of the dining philosophers (see Example 3.23) is, in dimensions 2 and 3:

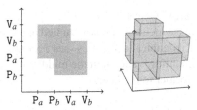

The geometric semantics introduced in this section bears many similarities with the semantics defined in the previous chapters, in particular with the cubical semantics introduced in Sect. 3.4: In some sense, under minor assumptions on the program, it is a continuous version of the cubical semantics, and reciprocally this explains why the elements of the cubical semantics could intuitively be seen as points, lines, surfaces, etc. This is formally described by the following theorem.

Proposition 4.19 *Suppose given a conservative program p satisfying the following condition: for every vertex x of the precubical set C_p the resource potential $\mathtt{r}$ is such that for every resource $a \in \mathcal{R}$, we have $\mathtt{r}(x)(a) \leqslant \kappa_a$, i.e., the program never releases more resources than the initial capacity. The geometric realization $\left|\check{C}_p\right|$ of its cubical semantics $\check{C}_p$ embeds, as a directed topological space, into its geometric semantics $\check{G}_p$.*

Proof The proof can roughly be done as follows. First, note that we have the isomorphism of d-spaces $|C_p| \cong G_p$: by induction, the isomorphism exists when p is reduced to an instruction, and the interpretations of other instructions correspond to each other (for instance we have $|C_{p\|q}| = |C_p \otimes C_q| = |C_p| \times |C_q| = G_p \otimes G_q$ because geometric realization sends tensor to cartesian product by Proposition 4.12). Since $\check{C}_p$ was obtained from C_p by restriction, it can be shown that $\left|\check{C}_p\right|$ can be obtained from $|C_p|$ by removing some subspace D. Finally, we have $|C_p| \setminus D = \left|\check{C}_p\right| \subseteq \check{G}_p = G_p \setminus R_p$ because $R_p \subseteq D$: the last inclusion follows from the fact that the resource potential from cubical semantics induces a potential $\mathtt{r}'$ on the geometric semantics which is always such that $\mathtt{r}(x)(a) \leqslant \mathtt{r}'(x)(a)$, and from the assumption on the resource potential of the program. $\qquad\square$

Remark 4.20 The embedding described in the previous proposition could be turned into an isomorphism by slightly modifying Definition 4.13 of geometric semantics and changing the definition for the resource functions of locking and unlocking to

$$r_{\mathrm{P}_a}(x)(b) = \begin{cases} 0 & \text{if } b \neq a \text{ or } x = 0 \\ -1 & \text{otherwise} \end{cases} \qquad r_{\mathrm{V}_a}(x)(b) = \begin{cases} 0 & \text{if } b \neq a \text{ or } x < 1 \\ 1 & \text{otherwise} \end{cases}$$

We chose not to define semantics in this way because it leads to much less readable figures. In fact the precise choice for coordinates does not really matter in the geometric model.

Example 4.21 (Swiss flag) Consider $p = (\mathrm{P}_a ; \mathrm{P}_b ; \mathrm{V}_b ; \mathrm{V}_a) \parallel (\mathrm{P}_b ; \mathrm{P}_a ; \mathrm{V}_a ; \mathrm{V}_b)$, where a and b are mutexes, which was already studied in Example 3.22. The directed geometric realization of its cubical semantics (see Example 3.40) is shown on the left and its geometric semantics is shown in the middle: it is easy to see that there is an embedding from the former to the later.

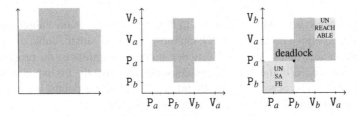

In the figure on the right, we have shown the point corresponding to the deadlock. Namely, note that the point with coordinates $(\mathrm{P}_b, \mathrm{P}_a)$ is the upper corner of a lower concavity of the state space: no future execution is allowed from this point, since increasing paths in the two coordinates would have to enter the forbidden region. This point thus corresponds to a deadlock. We have also drawn two regions depicting points which are, respectively, unsafe (i.e., might lead to a deadlock) and unreachable. This will be detailed in Sect. 4.2.

We will complete this comparison between the two models by showing in Theorem 4.38 that the directed paths in the geometric model correspond to paths in the cubical model. This will in particular allow us to define the semantics of a directed path as the semantics of the corresponding cubical path.

4.1.4 Simple Programs

In the following, in order to ease the presentation of some algorithms, we will often suppose that the programs we consider are in the simple form described below. This is of interest since their geometric semantics has a form which is particularly easy to manipulate.

Definition 4.22 A program is **simple** when it is of the form $p = p_1 \parallel p_2 \parallel \cdots \parallel p_n$, where the programs p_i are sequences consisting of actions, resource operations, and skip (i.e., the programs p_i do not contain conditional branching, loops, or parallel composition). In this case, the programs p_i are called the *processes* and n the *dimension* of the program.

Proposition 4.23 *The geometric semantics of a simple program $p = p_1 \parallel \cdots \parallel p_n$ is isomorphic to a d-space of the form*

$$\check{G}_p = \vec{I}^n \backslash \bigcup_{i=1}^{l} R^i \quad \text{with} \quad R^i = \prod_{j=1}^{n}]x_j^i, y_j^i[$$

with $l \in \mathbb{N}$ and, for every $i \in [1 : l]$ and $j \in [1 : n]$, $x_j^i, y_j^i \in \{-\infty\} \sqcup \vec{I} \sqcup \{\infty\}$ are such that $x_j^i < y_j^i$. The space R^i is called the ith forbidden region.

Example 4.24 The programs introduced in Example 4.15 (as well as Examples 4.18 and 4.21) are simple. For instance, with $\kappa_a = 1$, the program $\mathsf{P}_a ; \mathsf{V}_a \parallel \mathsf{P}_a ; \mathsf{V}_a$ has a geometric semantics which is isomorphic to $\vec{I}^2 \backslash R$ with $R =]\frac{1}{3}, \frac{2}{3}[\times]\frac{1}{3}, \frac{2}{3}[$. Again, the precise coordinates of the forbidden region R do not really matter since the properties we consider are up to isomorphism of d-spaces.

The regions R^i can be chosen so that each of them corresponds to a conflict on a particular resource a_i, in the expected way. It can be shown that if the capacity of a_i is κ_{a_i}, the region R^i is extended infinitely in at least $n - \kappa_{a_i} - 1$ directions, i.e., $x_j^i = -\infty$ and $y_j^i = \infty$ for at least $n - \kappa_{a_i} - 1$ values of $j \in [1 : n]$. For instance, the Swiss flag program in dimension 3 described in Example 4.45 has only mutexes and therefore the three forbidden regions are extended infinitely in one direction.

4.2 Homotopy in Directed Algebraic Topology

4.2.1 Classical Homotopy Theory

Algebraic topology is based on the notion of homotopy: two maps with the same domain and the same codomain are homotopic when one can be continuously deformed into the other. We only introduce the basic notions here, and the reader is referred to standard textbooks [4, 84] for a more detailed overview of the field. Possible generalizations of those notions to directed spaces are discussed in following sections.

Definition 4.25 Given two continuous maps $f, g : X \to Y$ between topological spaces, a **homotopy** from f to g is a continuous map $h : I \times X \to Y$ such that for every $x \in X$, $h(0, x) = f(x)$ and $h(1, x) = g(x)$. When there exists such a homotopy between f and g, the maps are said to be *homotopic*, which is denoted as $f \sim g$.

In the case where $X = I$, i.e., the maps f and g are paths in Y, such a homotopy is *endpoint-preserving* when for every $t, t' \in I, h(t, 0) = h(t', 0)$ and $h(t, 1) = h(t', 1)$.

Example 4.26 Consider the space $X = I \times I$. It is easy to show that any two paths $f, g : I \to X$ with same source and same target are necessarily homotopic. Namely, it can easily be checked that the function $h : I \times I \to X$ defined by $h(t, x) = (1 - t)f(x) + tg(x)$ is an endpoint-preserving homotopy between f and g.

In the following, when we consider a homotopy between two paths, we always implicitly assume that it is endpoint-preserving. A space is *simply connected* when it is path-connected and any two paths are homotopic.

One of the aim of algebraic topology is to classify *homotopy types* of spaces, i.e., spaces up to the following equivalence relation.

Definition 4.27 Two spaces X and Y are *homotopy equivalent* when there exists two continuous maps $f : X \to Y$ and $g : Y \to X$ such that $g \circ f \sim \mathrm{id}_X$ and $f \circ g \sim \mathrm{id}_Y$.

Intuitively, two homotopy equivalent spaces have the "same shape". For instance, a famous example is that a mug and a doughnut are homotopy equivalent. We will not discuss the generalizations of this notion in the directed setting and refer the reader to [77] for possible definitions.

4.2.2   Homotopy Between Directed Paths in Dimension 2

In order to provide concrete intuitions, we first consider homotopy between paths in geometric realizations of simple programs of dimension 2, which are subspaces of $\vec{I} \times \vec{I}$. As a first nontrivial example, consider the d-space $X = \vec{I} \times \vec{I} \setminus \{(x_0, y_0)\}$ for some point $(x_0, y_0) \in \vec{I} \times \vec{I}$. The following proposition characterizes when two directed paths are homotopic, i.e., we can continuously deform one path into the other, possibly going trough nondirected paths at intermediate stages. It follows directly from Theorem 4.47 which will be shown in Sect. 4.2.4.

Proposition 4.28 *Given two directed paths $f, g : I \to X$ with the same source and same target endpoints, the following are equivalent:*

(i) *the dipaths f and g are homotopic,*
(ii) *for every $t \in I, f(t) = (x_0, y)$ implies $y \gtrless y_0$ and $g(t) = (x_0, y)$ implies $y \gtrless y_0$, where $\gtrless$ is either $<$ or $>$,*
(iii) *there exists $t \in I$ such that $f(t) = (x_0, y)$ with $y \gtrless y_0$ and there exists $t' \in I$ such that $g(t') = (x_0, y)$ with $y \gtrless y_0$, where $\gtrless$ is either $<$ or $>$.*

Example 4.29 We have drawn the space $X = I \times I \setminus \{(x_0, y_0)\}$, on the left figure below.

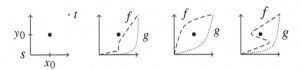

In the middle left are shown two homotopic directed paths. In particular, notice that the dashed one f is such that there are multiple instants $t \in I$ such that the first component of $f(t)$ is x_0. In the middle right, two nonhomotopic directed paths are shown. The figure on the right illustrates the importance of supposing that both of the paths are directed for (iii) to be equivalent to the others: here, the path f is not directed and not homotopic to g, however there exists $t \in I$ such that $f(t) = (x_0, y)$ with $y < y_0$ and similarly for g.

Given a simple two-dimensional program p, in the sense of Definition 4.22, the previous proposition can for instance be used—either directly or after showing easy variants left to the reader—to compute the homotopy classes of directed paths in its geometric semantics. In particular, the maximal paths (from s_p to t_p when the program is deadlock-free) are interesting because, as we will see in next section (Theorem 4.38), they correspond to executions of the program modulo commutation of actions.

Example 4.30 In the geometric semantics of the program $p = P_b \, ; V_b \, ; P_a \, ; V_a \, \| \, P_a \, ; V_a$, there are two maximal paths up to homotopy. For instance, the two directed paths above the hole are homotopic, whereas the path below is not homotopic to the two others:

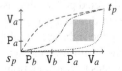

Example 4.31 The geometric realization of the programs

$$P_a \, ; V_a \, ; P_b \, ; V_b \, \| \, P_a \, ; V_a \, ; P_b \, ; V_b \qquad \text{and} \qquad P_a \, ; V_a \, ; P_b \, ; V_b \, \| \, P_b \, ; V_b \, ; P_a \, ; V_a$$

are, respectively.

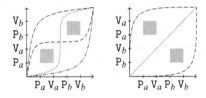

There are, up to homotopy with fixed endpoints, four maximal directed paths in the first d-space, whereas there are only three on the second d-space (we have shown one representative directed path for each homotopy class). Note however that their underlying topological spaces are homeomorphic: this shows that considering d-spaces instead of the underlying (nondirected) topological spaces often drastically changes the properties of the situation.

Lemma 4.32 *The homotopy relation satisfies the following properties:*

- *The relation $\sim$ is an equivalence relation.*
- *The relation $\sim$ is compatible with concatenation of paths: given four paths f, f' : $x \twoheadrightarrow y$ and $g, g' : y \twoheadrightarrow z$ in X, $f \sim f'$ and $g \sim g'$ implies $(f \cdot g) \sim (f' \cdot g')$.*
- *Concatenation of paths is associative and admits the empty path as identity up to homotopy: given paths $f : x \twoheadrightarrow y$, $g : y \twoheadrightarrow z$, and $h : z \twoheadrightarrow t$, we have $(f \cdot g) \cdot h \sim f \cdot (g \cdot h)$ and $\varepsilon_x \cdot f \sim f \sim f \cdot \varepsilon_y$.*

One can thus define the following category of paths up to homotopy in a given topological space; the previous lemma ensures that composition is well-defined and that the axioms of categories are satisfied.

Definition 4.33 The **fundamental groupoid** $\Pi_1(X)$ associated to a topological space X is the category whose objects are the points of X, and morphisms from x to y are equivalence classes of paths with x as source and y as target modulo endpoint-preserving homotopy. Composition of two paths is given by concatenation, and the identity on an object x is the equivalence class of the empty path on x.

For every path $p : I \to X$, the path $p^{-1}(t) = p(1-t)$ can be shown to provide an inverse for the corresponding morphism in the above category: every morphism is invertible, i.e., it is a *groupoid*. In the setting of directed spaces, the notion of endpoint-preserving homotopy between directed paths still makes sense, and one could define a category of homotopy classes of directed paths associated to a d-space. However, it turns out that this notion is not the right one: intuitively, this is because its allows continuously deforming a directed path into another by going through paths which might not be directed. This motivates the introduction in the next section of a directed variant of homotopy. Incidentally, the two notions of homotopy coincide in dimension 2, which is why no peculiarity could be observed in the examples up to now.

4.2.3 Dihomotopy and the Fundamental Category

Consider again the program $p = \mathrm{P}_a \,\text{;}\, \mathrm{V}_a \parallel \mathrm{P}_a \,\text{;}\, \mathrm{V}_a \parallel \mathrm{P}_a \,\text{;}\, \mathrm{V}_a$, where a is a resource of capacity 2, which has been used at the beginning of Sect. 3.4. Its cubical semantics is shown on the left (all the squares of length 2 are filled with 2-cubes) and its geometric semantics is the empty cube depicted on the right:

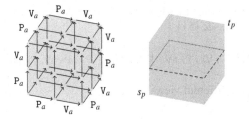

In the cubical semantics, consider the two horizontal paths in the middle, both labeled by $P_a . V_a . P_a . V_a$: the corresponding paths have been drawn in the geometric semantics. These two paths in the cubical semantics are not dihomotopic in the cubical sense (see Definition 3.65). This can easily be understood, keeping in mind that the resource a has initial capacity 2: after the "vertical" process has performed P_a, and before it performs V_a, the resource a has remaining capacity 1, and thus behaves as a mutex. We are thus essentially considering the program $P_a ; V_a \parallel P_a ; V_a$, which has two non-dihomotopic paths as extensively discussed in Chap. 3 (see Example 3.39 in particular). Note however that in the geometric semantics, the two paths *are homotopic*. We can continuously deform one path into the other by going below the cube for instance, as shown in the following "movie":

Some of the paths used during this homotopy are not directed, and it is easy to see that it is not possible to construct a homotopy between the two paths which would only go through directed paths. Intuitively, this means that for this equivalence to be true, one has to be able to execute some actions "backwards in time," which is counter to the interpretation in terms of program execution. In order for the equivalence between paths in the geometric model and the cubical model to match, we therefore investigate a directed variant of the notion of homotopy.

Definition 4.34 Given a d-space X, a **dihomotopy** between two directed paths $f, g :$ $\vec{I} \to X$ is an endpoint-preserving homotopy $h : I \times \vec{I} \to X$ from f to g such that for every $t \in [0, 1]$, the path $x \mapsto h(t, x)$ is directed. In this case the paths f and g are called *dihomotopic*, and denoted as $f \sim g$.

Note that, by Proposition 4.9, a dihomotopy can also be considered as a morphism $I \to X^{\vec{I}}$, i.e., a nondirected path in the directed path space of X. Obviously, two dihomotopic paths are homotopic, but the contrary is not necessarily true, as explained above and illustrated in the following examples.

Example 4.35 (*Room with three barriers*) The geometric semantics associated to the following program, see [51],

$$\mathbb{P}_a ; \mathbb{V}_a ; \mathbb{P}_b ; \mathbb{V}_b ; \mathbb{P}_c ; \mathbb{V}_c \ \| \ \mathbb{P}_b ; \mathbb{V}_b \ \| \ \mathbb{P}_a ; \mathbb{P}_b ; \mathbb{V}_a ; \mathbb{P}_c ; \mathbb{V}_b ; \mathbb{V}_c$$

is shown below (with different views):

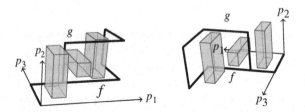

where a and c are mutexes, and b is a resource of capacity 2. We call the three processes p_1, p_2, and p_3, so that the above program is $p_1 \| p_2 \| p_3$ in that order. The first dipath f corresponds to the case where p_3 locks a before p_1 and then p_2 locks b after p_1 and p_3 have locked b together and released it, i.e., to the following trace, where superscripts indicate the number of the process performing the action:

$$f \ = \ \mathbb{P}_a^3 \cdot \mathbb{P}_b^3 \cdot \mathbb{V}_a^3 \cdot \mathbb{P}_a^1 \cdot \mathbb{V}_a^1 \cdot \mathbb{P}_b^1 \cdot \mathbb{V}_b^1 \cdot \mathbb{P}_c^1 \cdot \mathbb{V}_c^1 \cdot \mathbb{P}_c^3 \cdot \mathbb{V}_b^3 \cdot \mathbb{V}_c^3 \cdot \mathbb{P}_b^2 \cdot \mathbb{V}_b^2$$

The second dipath g corresponds to the situation where p_3 locks a before p_1 and then p_1 locks b after p_2 and p_3 have locked b together and release it:

$$g \ = \ \mathbb{P}_a^3 \cdot \mathbb{P}_b^3 \cdot \mathbb{V}_a^3 \cdot \mathbb{P}_a^1 \cdot \mathbb{V}_a^1 \cdot \mathbb{P}_b^2 \cdot \mathbb{V}_b^2 \cdot \mathbb{P}_b^1 \cdot \mathbb{V}_b^1 \cdot \mathbb{P}_c^1 \cdot \mathbb{V}_c^1 \cdot \mathbb{P}_c^3 \cdot \mathbb{V}_b^3 \cdot \mathbb{V}_c^3$$

These two dipaths are homotopic, since one can deform continuously one into another through paths going around the central hole. However, some of the paths in between are necessarily nondirected and the two paths are not dihomotopic.

Example 4.36 (*Two wedges*) Consider the following program:

$$\mathbb{P}_c ; \mathbb{P}_a ; \mathbb{P}_b ; \mathbb{V}_b ; \mathbb{P}_d ; \mathbb{V}_d ; \mathbb{V}_c ; \mathbb{V}_a \ \| \ \mathbb{P}_b ; \mathbb{P}_d ; \mathbb{P}_a ; \mathbb{V}_a ; \mathbb{P}_c ; \mathbb{V}_c ; \mathbb{V}_d ; \mathbb{V}_b$$
$$\| \ \mathbb{P}_a ; \mathbb{P}_b ; \mathbb{V}_a ; \mathbb{V}_b ; \mathbb{P}_c ; \mathbb{P}_d ; \mathbb{V}_c ; \mathbb{V}_d$$

where a, b, c, and d are resources of capacity 2, and where, once again, we call the first, second, and third processes, p_1, p_2, and p_3, respectively, from left to right. Its geometric semantics has two wedges as forbidden region:

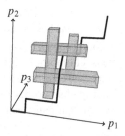

All dipaths from beginning to end are homotopic, including the one pictured above, which corresponds to the execution trace below:

$$P_c^1 . P_b^2 . P_a^1 . P_b^1 . V_b^1 . P_d^2 . P_a^2 . V_a^2 . P_a^3 . P_b^3 . V_a^3 . V_b^3 . P_c^3 . P_d^3 . V_c^3 . V_d^3 . P_d^1 . V_d^1 . P_c^2 . V_c^2 . V_c^1 . V_a^1 . V_d^2 . V_b^2$$

However, not all d-paths are mutually *di*homotopic: there are two dihomotopy classes, one with paths going "in the middle" as the above one, and one going outside the two wedges.

The dihomotopy relation satisfies the same properties as the homotopy relation listed in Lemma 4.32, and we can therefore, as before, define a category as follows.

Definition 4.37 The **fundamental category** $\vec{\Pi}_1(X)$ of a d-space X is the category whose objects are the points of X and whose morphisms are the directed paths up to dihomotopy.

In Proposition 4.19, we saw that given a program p, its geometric semantics $\check{G}_p$ corresponds to its cubical semantics $\check{C}_p$ via the directed geometric realization: formally, $\check{G}_p = \left| \check{C}_p \right|$. In particular, to each vertex x in $\check{C}_p$ corresponds a point $|x|$ in $\check{G}_p$. We are now ready to show that both semantics essentially have the same (directed) paths, i.e., model the execution of programs in the same way. Of course there are many more paths in the geometric model, so this correspondence will only be valid up to dihomotopy. It would not be valid up to homotopy, which makes another strong argument in favor of dihomotopy as the right notion of equivalence in the context of directed spaces.

Theorem 4.38 *Consider a conservative program p with $\check{C}_p$ as cubical semantics and $\check{G}_p$ as geometric semantics. Given two vertices x and y there is a bijection between paths in $\check{C}_p$ from x to y up to dihomotopy (in the sense of cubical sets) and paths in $\check{G}_p$ from $|x|$ to $|y|$ up to dihomotopy. Equivalently, the functor induced by the directed geometric realization*

$$\vec{\Pi}_1\left(\check{C}_p\right) \hookrightarrow \vec{\Pi}_1\left(\check{G}_p\right)$$

between the fundamental category of $\check{C}_p$ (see Definition 3.36) and the fundamental category of $\check{G}_p$ (see Definition 4.37) is full and faithful.

Proof Starting from two dipaths $f, g : x \twoheadrightarrow y$ in $\check{C}_p$, whose realizations in $\check{G}_p$ are dihomotopic dipaths $|f|, |g| : |x| \twoheadrightarrow |y|$, one has to show that f and g are dihomotopic in $\check{C}_p$. The idea is to start from the dihomotopy $h : I \to X^I$ from $|f|$ to $|g|$, and to show that at each time $t \in I$, the path $h(t)$ is dihomotopic to a dipath which is the realization of a dipath in $\check{C}_p$. A standard compactness argument shows that one only needs to consider a finite number of cubical dipaths in order to cover all the dipaths $h(t)$ for $t \in I$, up to dihomotopy, and those cubical paths can be shown to be dihomotopic. Details can be found in [44].   □

Example 4.39 Consider the program $P_b ; V_b ; P_a ; V_a \parallel P_a ; V_a$ already seen in Example 4.30. The cubical semantics is shown on the left and the geometric semantics on the right. The dotted directed path in the geometric semantics is dihomotopic to the dashed path which is the image of a path in the cubical semantics:

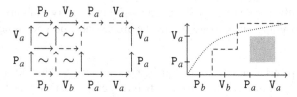

At first, it might seem that the previous theorem could be extended in order to show that there is an equivalence between the fundamental categories (or even that the two spaces have the same "directed homotopy type"). This is not the case, and a suitable categorical notion of equivalence adapted to this situation is quite subtle [77]. For instance, if one considers a portion of the dotted path in the above example, no dihomotopic path which is the image of a cubical path can be found. However, one can always extend a path so that it is dihomotopic to the image of a cubical path:

Proposition 4.40 *Given a conservative program p, for every path $f' : x' \twoheadrightarrow y'$ in $\check{G}_p$ there exists a path $f : x \twoheadrightarrow y$ in $\check{C}_p$ and paths $f'_1 : |x| \twoheadrightarrow x'$ and $f'_2 : y' \twoheadrightarrow |y|$ such that the path $f'_1 . f' . f'_2$ and $|f|$ are dihomotopic:*

$$
\begin{array}{ccc}
 & \overset{f'}{x' \longrightarrow y'} & \\
\overset{f'_1}{\nearrow} & & \overset{f'_2}{\searrow} \\
|x| \cdots\cdots\cdots\cdots\cdots\cdots\cdots\cdots\cdots\cdots\cdots\longrightarrow |y| \\
 & |f| &
\end{array}
$$

Moreover, if x' (resp. y') is the realization of a point in $\check{C}_p$ then x (resp. y) can always be chosen to be this point.

Remark 4.41 The extensions f'_1 and f'_2 provided in the previous proposition are not canonical. For instance, in the geometric semantics of $P_a ; V_a \parallel P_a ; V_a$ depicted on the left, there are two possible extensions on the right f'_2 and f''_2 of the path f':

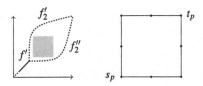

On the right is depicted the geometric realization of $\check{C}_p$ (the interior of the square is not filled), which has two maximal paths (from s_p to t_p) which are, respectively, dihomotopic to $f_1' \cdot f' \cdot f_2'$ and $f_1' \cdot f' \cdot f_2''$.

We can now define the semantics of a path by reusing the semantics developed in Chap. 3 as follows.

Definition 4.42 Given a coherent conservative program p and two vertices $x, y \in \check{C}_p$, the *operational semantics* of a path $t : |x| \twoheadrightarrow |y|$ is the function $[\![t]\!] : \Sigma \to \Sigma$ defined as $[\![u]\!]$ (in the sense of Definition 2.18 and Sect. 3.1.2) for some path $u : x \twoheadrightarrow y$ in $\check{C}_p$ such that $|u| = t$.

The existence of such a path u is granted by Theorem 4.38. Moreover, the definition does not depend on the choice of the path u. Namely, if we have two paths $u, u' : x \twoheadrightarrow y$ such that $|u| = |u'| = t$, then by Theorem 4.38 we have $u \sim u'$, and therefore $[\![u]\!] = [\![u']\!]$ because the program is supposed to be coherent. It can be shown that extremal endpoints (such as s_p, t_p, deadlocks, etc.) are the geometric realization of a vertex in $\check{C}_p$, so the above definition covers the majority of interesting paths (in particular maximal execution traces $t : s_p \twoheadrightarrow t_p$). This definition does not extend easily to other paths since the extension given by Proposition 4.40 is not canonical in general as noted in Remark 4.41. However, a semantics can actually be associated to all paths by adapting the definitions elaborated in Sect. 2.2 to the geometrical setting, i.e., we can formally define the "effect" of a geometric path on a state. We did not do this here to avoid redundancies.

Remark 4.43 Consider a coherent program of the form $p = A \parallel B$. Its cubical semantics is shown on the left and its geometric semantics on the right:

$$\check{C}_p \;=\; B \!\!\begin{array}{c} \xrightarrow{\;\;A\;\;} \\ \sim \\ \xrightarrow[\;\;A\;\;]{} \end{array}\!\! B \qquad \check{G}_p \;=\; B \!\!\begin{array}{c} {}^{t} \\ \\ \xrightarrow[\;\;A\;\;]{} \end{array}$$

The dotted path t is dihomotopic to the geometric realization of the two maximal paths, respectively, labeled by $A \cdot B$ and $B \cdot A$ in $\check{C}_p$, so that its semantics is $[\![A \cdot B]\!] = [\![B \cdot A]\!]$. We can see here the importance of supposing the program to be coherent: for instance, with $A = \mathtt{x:=1}$ and $B = \mathtt{x:=2}$, the program would not be coherent, and indeed the semantics would clearly not be well-defined.

The previous propositions also allow us to consider the following "undesirable points," which are the geometric counterparts of the positions introduced in Definition 2.28 and Proposition 3.19:

Definition 4.44 Given a conservative program p, we can identify the following points in its geometric semantics $\check{G}_p$.

- A point x such that there is no dipath $t : s_p \twoheadrightarrow x$ is *unreachable*.
- A point x different from t_p and such that the only dipath from x is the constant dipath ε_x is a *deadlock*.
- A point x such that there is a dipath $t : x \twoheadrightarrow y$ for some deadlock y is *unsafe*.
- A point x such that there is no dipath $t : x \twoheadrightarrow t_p$ is *doomed*.

The subspace of $\check{G}_p$ consisting of points which are unreachable (resp. unsafe, resp. doomed) is called the *unreachable region* (resp. the *unsafe region*, resp. the *doomed region*). Given a deadlock y, the set of points x such that every maximal path originating in x has y as target is called the doomed region associated to the deadlock y.

The unsafe and unreachable regions are illustrated in Example 4.21, for the Swiss flag program.

4.2.4 Simple Programs with Mutexes

It was noted in Sect. 4.2.2 that homotopy seemed to be a reasonable notion for two-dimensional simple programs. The explanation is that in this case, it coincides with dihomotopy. We show here that this is the case for two-dimensional programs, and more generally for programs with resources of capacity 1 (i.e., mutexes) only.

Suppose given a simple program p of given dimension n such that all the resources used in the program are mutexes. We write $X = \check{G}_p$ for its geometric semantics. Recall from Proposition 4.23 that X is (up to isomorphism) a subspace of $\vec{I}^n$: it is of the form $\vec{I}^n \setminus \bigcup_{i=1}^{l} R^i$ where the R^i are hypercubes. Moreover, since mutexes are used, these hypercubes are infinitely extended in $n - 2$ dimensions.

Example 4.45 (*3d Swiss flag*) Consider the program $p = \mathrm{P}_a \,;\mathrm{V}_a \parallel \mathrm{P}_a \,;\mathrm{V}_a \parallel \mathrm{P}_a \,;\mathrm{V}_a$ with $\kappa_a = 1$. Its geometric semantics is $\check{G}_p = \vec{I}^3 \setminus \left(R^1 \cup R^2 \cup R^3 \right)$ with the region $R^1 =]-\infty, \infty[\times]\frac{1}{3}, \frac{2}{3}[\times]\frac{1}{3}, \frac{2}{3}[$; the other two forbidden regions R^i are obtained by symmetry:

Given points x and y, we write x_i for the ith coordinate of x (and similarly for vectors) and $y - x$ for the vector from x to y. We will consider the elements of X as partially ordered by the *product order*: given two points x and y, we have $x \leqslant y$ when $x_i \leqslant y_i$ for every index i. Two points x and y such that neither $x \leqslant y$ nor $y \leqslant x$ are said to be *achronal*: in particular, any two points in a directed path in X are not achronal. The poset $(\vec{I}^n, \leqslant)$ is a lattice with $(x \vee y)_i = \max(x_i, y_i)$ and $(x \wedge y)_i = \min(x_i, y_i)$. Given two points x and y, we write $x \square y = \{z \in X \mid x \wedge y \leqslant z \leqslant x \vee y\}$ for the *hypercube* generated by x and y. This structure can be used to show the following simple lemma, whose proof technique can often be used in order to construct dihomotopies.

Lemma 4.46 *Suppose given two directed paths $f, g : x \rightarrowtail y$ in X with the same source and the same target. There exists directed paths f' and g' with the same image as f and g, respectively, such that for every $t \in I$, the points $f'(t)$ and $g'(t)$ are either achronal or equal.*

Proof If $x = y$, the paths f and g are constant and $f' = f$ and $g' = g$ satisfies the conditions. Now, assume $x \neq y$. We write v for the vector $y - x$ and d for the straight line from x to y, defined by $d(t) = x + tv$. Since the paths f and g are directed, we have $v_i \geqslant 0$ for every index i. Given $t \in I$, we also write H_t for the space orthogonal to d going through the point $d(t)$: $H_t = \{z \in X \mid (z - d(t)) \cdot v = 0\}$. The space H_t intersects the image of f (and similarly for g) in exactly one point. Namely, suppose that z and z' are two distinct points in their intersection, if $v_i = 0$ then $z_i = z'_i = x_i = y_i$ and therefore, since the v_i are positive and $(z - z') \cdot v = 0$, there exists j and k such that $z_j - z'_j > 0$ and $z_k - z'_k < 0$. Since both points belong to f, this contradicts the fact that f is directed.

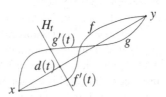

We define $f'(t)$ (resp. $g'(t)$) as the intersection point of H_t with the image of f (resp. g), and the paths thus defined are suitable for similar reasons as before: the coordinates of the vector $g'(t) - f'(t)$ cannot be all of the same sign (if the vector is not null) because both $f'(t)$ and $g'(t)$ belong to H_t. □

In the previous lemma, it can be noted that the path f' we constructed is injective when $x \neq y$. It is therefore bijective on its image: this inverse is continuous because I is compact, and preserves the order because I is totally ordered. Therefore, f can be obtained from f' by a directed reparametrization: we have $f = f' \circ \theta$ with $\theta = f'^{-1} \circ f$. From this it follows easily that the maps f and f' are dihomotopic and similarly $g \sim g'$. We can therefore use the preceding lemma to show that homotopy and dihomotopy coincides in the space X.

Theorem 4.47 *Given two dipaths f and g in X, the following are equivalent:*

(i) *for every pair of achronal points x and y and in the image of f and g, respectively, we have $x \square y \subseteq X$,*

(ii) *the paths f and g are dihomotopic,*

(iii) *the paths f and g are homotopic.*

Proof Suppose (i) verified and apply Lemma 4.46. Given $t \in I$, the points $f'(t)$ and $g'(t)$ are either achronal or equal, and therefore by (i) the straight line from one to the other belongs to the space. The map $h : I \to X^I$ defined by $h(t) = (1 - t)f' + tg'$ is thus a well-defined dihomotopy from f' to g' and (ii) is satisfied. Moreover, two dihomotopic paths are necessarily homotopic, i.e., (ii) implies (iii). Finally, suppose (iii) is verified and write h for a homotopy between f and g. Now, take two achronal points x and y in the image of f and g, respectively, and suppose that a point $z \in x \square y$ does not belong to X.

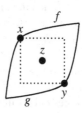

Now, suppose that we are dimension 2 (the general case will be deduced from this one). Since f and g are increasing, it is easily shown that one of the two paths goes "above" z and the other one goes "below" z. The nondirected path $f \cdot \overline{g}$ is thus winding around z, and the homotopy h would induce a homotopy between $f \cdot \overline{g}$ and $g \cdot \overline{g}$ in $\overline{I}^2 \setminus \{z\}$, where $\overline{g}$ is the nondirected path g taken backward. Since $g \cdot \overline{g}$ is homotopic to the constant path this is absurd, because $\overline{I}^2 \setminus \{z\}$ is homotopy equivalent to a circle. If we do not suppose that we are in dimension 2, we know that z belongs to a region R^i which is infinitely extended in all dimensions but two. We can reach a contradiction as before by keeping R^i as the only forbidden region and projecting on the plane given by those two dimensions: the form of R^i ensures that if there is a homotopy in the original space, there is one in the projected space. □

The geometric semantics of a simple program of dimension 2 can be shown to be isomorphic to the geometric semantics of one using only mutexes: intuitively, since there are only two processes, it is enough to have binary synchronizations. Therefore, we have as a direct corollary,

Proposition 4.48 *In the geometric semantics of a simple program of dimension 2, two paths are homotopic if and only if they are dihomotopic.*

This explains why the case of dimension 2, and more generally for programs with only mutexes, is quite simple compared to the unrestricted case where the above proposition does not hold as explained in Sect. 4.2.3.

The previous theorem has a number of other interesting consequences. If we write $U : \mathbf{dTop} \to \mathbf{Top}$ for the forgetful functor, the obvious functor $\vec{\Pi}_1(X) \to \Pi_1(UX)$ is faithful, and therefore every morphism is both epi and mono in $\vec{\Pi}_1(X)$ because it is the case in $\Pi_1(UX)$ since that is a groupoid. Notice that those properties are quite similar to those shown in the algebraic counterpart in Sect. 3.3.4. Finally, difficult algorithmic problems such as finding homotopy classes of paths can be efficiently addressed in dimension 2 [152].

4.2.5 D-Homotopy

In this section, we investigate an alternative definition for directed homotopy. While it provides a different equivalence than dihomotopy in general, it coincides with dihomotopy in geometric semantics of programs. A homotopy $h : I \times I \to X$ between two paths in a topological space X can be seen as a path $h : I \to X^I$ in the space X^I of paths in X. It is thus natural to investigate a directed variant of this notion, usually called d-homotopy.

Definition 4.49 Given a d-space X and two paths $f, g \in X^{\vec{I}}$ with same source and same target, an *elementary d-homotopy* between f and g is a d-map $h : \vec{I} \to X^{\vec{I}}$ such that $h(0) = f$, $h(1) = g$. It is endpoint-preserving when for every $t \in \vec{I}$, $h(t)(0) = f(0) = g(0)$ and $h(t)(1) = f(1) = g(1)$.

For two paths the relation of being related by a d-homotopy is not an equivalence relation in general. We thus define:

Definition 4.50 The *d-homotopy* relation on paths of a given d-space X is the smallest equivalence relation $\sim_d$ such that $f \sim_d g$ whenever there exists an elementary endpoint-preserving d-homotopy from a path f to a path g in X.

It can easily be shown that two d-homotopic paths are necessarily dihomotopic, but the converse is not true:

Example 4.51 Consider the 2-sphere S^2 equipped with the d-space structure such that directed paths are those with increasing latitude and constant longitude. Then all d-paths from the south pole to the north pole are dihomotopic but two d-paths are d-homotopic if and only if they share the same longitude.

However, the geometric models of concurrent programs are geometric realizations of precubical sets satisfying particular properties (they are so-called *geometric* precubical sets), for which d-homotopy and dihomotopy coincide [44]. We will thus only consider dihomotopy in the following.

4.3 Constructions on the Fundamental Category

We provide, in this section, the extension to the directed setting of some classical constructions in algebraic topology on the fundamental category. These directed counterparts are however much less well established and some details of their axiomatization are still under debate. Another very important construction on the fundamental category is the category of components, which will be described and investigated in Chap. 6.

4.3.1 The Seifert–Van Kampen Theorem

The so-called "Seifert–van Kampen theorem" is a celebrated result in algebraic topology [84]: it enables one to compute the fundamental group of a space by knowing the fundamental group of some suitable subspaces, thus providing some form of modularity in the computation of fundamental groups. It has since then been generalized to the computation of fundamental groupoids [19]. We describe here some of its generalizations to the setting of d-spaces, as a tool to compute fundamental categories [66, 76].

Theorem 4.52 *Suppose given a d-space X together with two subspaces Y and Z such that the union of their interiors covers X: the diagram on the left, whose morphisms are inclusion maps, is a pushout in* **dTop**. *Then its image, drawn on the right, is a pushout in* **Cat**.

$$
\begin{array}{ccc}
Y \cap Z & \longrightarrow & Z \\
\downarrow & & \downarrow \\
Y & \longrightarrow & X
\end{array}
\qquad\qquad
\begin{array}{ccc}
\vec{\Pi}_1(Y \cap Z) & \longrightarrow & \vec{\Pi}_1(Z) \\
\downarrow & & \downarrow \\
\vec{\Pi}_1(Y) & \longrightarrow & \vec{\Pi}_1(X)
\end{array}
$$

Example 4.53 Consider the space $X = \vec{S}^1$ the directed circle, see Example 4.4, whose points are elements of the complex plane of the form $z_\theta = e^{i\theta}$ with $\theta \in \mathbb{R}$. Write Y (resp. Z) for the set $\vec{S}^1 \setminus \{e^{i0}\}$ (resp. $\vec{S}^1 \setminus \{e^{i\pi}\}$). The categories $\vec{\Pi}_1(Y)$ and $\vec{\Pi}_1(Z)$ are both isomorphic to the category generated by the poset $]0, 1[$. A computation of the pushout in **Cat** shows that the monoid $\vec{\Pi}_1(\vec{S}^1)(x, x)$ is isomorphic to $(\mathbb{N}, +, 0)$, and the whole category can be characterized in a similar way.

Example 4.54 The Seifert–van Kampen theorem can be used to compute the fundamental category of any graph. For instance, consider the graph G pictured on the left:

Its directed geometric realization $|G|$ can be obtained as a colimit of the four open directed topological spaces shown on the right, an edge not ending with a vertex being open. In a more general way, a (directed geometric realization of a) finite graph is finitely covered by open stars around each vertex, an open star being a neighborhood of a vertex that contains only one vertex. Theorem 4.52 can then be applied, on noting that the fundamental category of an open star containing a vertex x is isomorphic to the poset obtained by gluing (at the origin) n copies of $(\mathbb{R}_-, \leqslant)$ and p copies of $(\mathbb{R}_+, \leqslant)$ at 0, where n (resp. p) is the number of ingoing (resp. outgoing) arrows of the vertex x.

4.3.2 The Universal Dicovering Space

Many of the practical and theoretical constructions developed in this book assume that the programs we consider are *loop-free* (this is in particular the case for simple programs): we suppose that they do not contain `while` constructions, or equivalently that the corresponding geometric semantics do not contain nontrivial directed loops, i.e., the associated fundamental category is loop-free, see Chap. 6. In order to remove this limitation, one is interested in theoretically *unrolling* programs with loops, by unfolding the loops and convert them into (potentially infinite) programs without loops.

A syntactical way of performing this is to consider the contextual equivalence

$$\texttt{while } b \texttt{ do } c \quad \approx \quad \texttt{if } b \texttt{ then}(c;\texttt{while } b \texttt{ do } c)\texttt{else skip}$$

presented in Remark 2.26: by applying this equivalence, from left to right, an infinite number of times, a `while` loop can be replaced by an infinite sequence of conditional branching, at the cost of considering infinite programs, or having a bound on the length of considered executions. Such an approach can be formalized, and is actually traditional in denotational semantics and verification. We present here a more geometrical approach to this construction, based on universal covering spaces for directed spaces.

The universal covering space of a topological space is well known and a widely used construction in traditional algebraic topology: in some sense, it provides an "unfolded" version of a space in the sense of unfolding loops used here. We investigate a directed variant of this construction, called the *universal dicovering space*, introduced in [43, 46], which should fit the following specification: starting from a d-space, it should produce another d-space, whose dipaths are suitably related with those of the original space, which contains no nontrivial (directed) loops, or at least no loops which are reachable from the start point in the case of the geometric semantics of a program. Moreover, this construction should be conservative in the sense that if we start from a (nondirected) topological space and see it as a d-space (with all paths being directed) the universal dicovering space should be the universal covering space of the topological space. We begin by briefly recalling some classical definitions in the nondirected case, a more detailed presentation can be found in most algebraic topology textbooks [84].

Definition 4.55 A continuous map $p : Y \to X$ between path-connected spaces X and Y is a *covering* if for every point $x \in X$ there exists a neighborhood U of x and a set F such that $p^{-1}(U)$ is a disjoint union of a family of open sets $(U_i)_{i \in F}$ of Y and for each $i \in F$ the restriction $p_i : U_i \to U$ of p is a homeomorphism. In this situation, X is called the *base space* and Y a *covering space*. A morphism between two coverings $p_1 : Y_1 \to X$ and $p_2 : Y_2 \to X$ of the same base space X is a continuous map $f : Y_1 \to Y_2$ such that $p_1 = p_2 \circ f$.

These spaces enjoy quite nice properties, in particular the paths in the base space have a counterpart in the covering space, which can be formalized as follows:

Definition 4.56 Given continuous maps $f : X_1 \to X_2$ and $g : Y_1 \to Y_2$ between topological spaces, g has the *unique right lifting property* with respect to f if for every pair of continuous maps $h_1 : X_1 \to Y_1$ and $h_2 : X_2 \to Y_2$ such that $g \circ h_1 = h_2 \circ f$, there exists a unique map $h : X_2 \to Y_1$ such that $h \circ f = h_1$ and $g \circ h = h_2$:

$$
\begin{array}{ccc}
X_1 & \xrightarrow{\;h_1\;} & Y_1 \\
\scriptstyle f \downarrow & \nearrow^{\scriptstyle h} & \downarrow \scriptstyle g \\
X_2 & \xrightarrow[\;h_2\;]{} & Y_2
\end{array}
$$

In particular, when $f : \{0\} \to I$ is the inclusion of the one-point space into the interval, we say that g has the *unique path lifting property*. More generally, when g has the unique right lifting property w.r.t. the inclusion map $f : X \times \{0\} \to X \times I$ for any topological space X, we say that g has the *unique homotopy lifting property*.

Proposition 4.57 *Every covering map $p : Y \to X$ has the unique homotopy lifting property (in particular, it has the unique path lifting property).*

For a reasonable class of spaces, covering maps can actually be characterized as maps having suitable unique lifting properties [18].

We now introduce universal covering spaces; it is advisable to do so in the setting of pointed spaces.

Definition 4.58 A *pointed space* (X, x) is a pair consisting of a space X and a *base point* $x \in X$, and a *pointed morphism* $f : (X, x) \to (Y, y)$ is a continuous map $f : X \to Y$ such that $y = f(x)$.

A covering space of a space X can be thought of as the space X in which some loops have been unrolled, and it is natural to consider a "maximal" such space, in which all loops have been unrolled. A pointed morphism $p : (Y, y) \to (X, x)$ such that $p : Y \to X$ is a covering called a *pointed covering*.

Definition 4.59 A pointed covering $p : (\tilde{X}, \tilde{x}) \to (X, x)$ is a *universal covering* if for every pointed covering $q : (Y, y) \to (X, x)$, there exists a unique pointed morphism $f : (\tilde{X}, \tilde{x}) \to (Y, y)$ satisfying $p = q \circ f$.

Remark 4.60 If we write **pTop** for the category of pointed spaces and pointed morphisms, the category of pointed coverings of a pointed space (X, x) can be seen as a full subcategory of the slice category **pTop**$/(X, x)$. A universal covering of (X, x) is precisely an initial object in this category.

When the space X is *well-connected* (i.e., path-connected, locally path-connected, and semilocally simply connected), a universal covering $(\tilde{X}, \tilde{x})$ of (X, x) exists; it is uniquely defined up to covering isomorphism, and does not actually depend on the choice of the base point $x \in X$ up to homeomorphism: we thus simply refer to the universal covering $\tilde{X}$ of X. The space $\tilde{X}$ can be shown to be simply connected, i.e., all loops are contractible, and $(\tilde{X}, \tilde{x})$ can actually be characterized as the simply connected pointed covering of (X, x).

Example 4.61 The map $p : (\mathbb{R}, 0) \to (S^1, 1)$ given by $p(t) = e^{2i\pi t}$ is a universal covering:

Given a path $f : I \to S^1$ and a point $x \in \mathbb{R}$ such that $p(x) = f(0)$, there exists a unique path $g : I \to \mathbb{R}$ such that $g(0) = x$ and $p \circ g = f$. The paths in $\mathbb{R}$ can be thought of as "unrollings" of paths in S^1: a path $f : I \to \mathbb{R}$ of the form $f(t) = kt$ for some $k \in \mathbb{R}$ maps to a path which loops the circle k times, and conversely paths in S^1 which loop k times lift to paths of length k.

Example 4.62 The map $q : (S^1, 1) \to (S^1, 1)$ given by $p(e^{2i\pi t}) = e^{8i\pi t}$ is a covering, but not the universal one, in which every fiber $p^{-1}(x)$ has four elements. Note that a path in the base space looping less than 4 times lifts to a nonlooping path, but a path looping four times loops the covering only once: some, but not all, loops are unrolled. The unique map from the universal covering is $f(t) = e^{i\pi\frac{t}{2}}$.

Remark 4.63 The above definitions of coverings and universal coverings would still make sense if we do not assume that spaces are path-connected, but this generality gives rise to situations which are less commonly encountered in topology. The universal pointed covering of a pointed space (X, x) which is not path-connected would be the one defined as usual above over the path-connected component containing the base point, and with empty fibers above other path components, reflecting the fact that the empty set is an initial object in the unpointed slice category **Top**$/X$. In this case, the universal covering would depend on which path component contains the base point, but not on the choice within the component.

The universal covering space can be explicitly described as follows:

Proposition 4.64 *Given a pointed topological space (X, x), which is well connected, the universal covering space exists and is isomorphic to the pointed space $(\tilde{X}, \tilde{x})$ whose points are the equivalence classes $[f]$ under endpoint-preserving homotopy of paths $f : x \twoheadrightarrow y$ in X starting from x, with base point $\tilde{x}$ the class of the constant path $x \twoheadrightarrow x$, and covering map $p : \tilde{X} \to X$ the endpoint map defined by $p([f]) = f(1)$. The topology on $\tilde{X}$ is generated by sets $U_{[f]}$, where $U \subseteq X$ is an open set and $[f] \in \tilde{X}$ is the homotopy class of paths in X such that $f(1) \in U$, defined by $U_{[f]} = \{[f \cdot g] \mid g : I \to U, g(0) = f(1)\}$.*

Remark 4.65 This construction is generalized to the directed case in Proposition 4.75, which is why we stress some properties that are used here. The topology on $\tilde{X}$ clearly implies that the map $p : (\tilde{X}, \tilde{x}) \to (X, x)$ defined above is continuous. Moreover, given a covering $q : (Y, y) \to (X, x)$, the unique pointed function $r : (\tilde{X}, \tilde{x}) \to (Y, y)$ such that $q \circ h = p$ is determined by the unique path lifting property: we have $r([f]) = \hat{f}(1)$ where $\hat{f} : X \to Y$ is the unique lift of f in Y with initial point y given by Proposition 4.57. The reason for requiring the space X to be well connected is, that otherwise, p may not be a covering and/or r may not be continuous. For the universal dicovering, Proposition 4.75, the maps p and r are defined as above, but much less is required of the topology on X: a dicovering is defined by lifting properties and in particular it does not have to satisfy the local "layering" properties of a classical covering. Hence, universal dicoverings exist for a much larger class of d-spaces Proposition 4.75; in fact, there is a universal dicovering of every pointed d-space. But the topology on the universal dicovering has to be made Δ-generated, see Definition 4.73.

The notion of a (universal) covering space can be generalized in the directed setting as follows [43, 45, 46, 53]. Instead of mimicking the classical definition of coverings (see Remark 4.82), this generalization focuses on the lifting (and "unrolling") properties, see Proposition 4.57. It follows the intuition that, in the geometric semantics of a program p, we only need to lift paths which correspond to execution traces, i.e., those which are starting from the beginning point s_p. Moreover, a nice outcome of this approach is that the resulting space can be described using a variant of the construction given in Proposition 4.64, which is very natural from a computer-scientific point of view. We will use the notion of (pointed) d-space, which is an immediate extension of the one in the nondirected setting (see Definition 4.58). In the following, the geometric semantics of a program will implicitly be considered as a pointed space, with its beginning point as base point.

Definition 4.66 Given d-spaces X and Y, a morphism $p : Y \to X$ is a **dicovering** if it has the unique right lifting property w.r.t. the inclusion d-maps

$$\{0\} \hookrightarrow \vec{I} \qquad \{(0, 0)\} \hookrightarrow (I \times \vec{I})/\approx_0 \qquad \{(0, 0)\} \hookrightarrow (I \times \vec{I})/(\approx_0 \cup \approx_1)$$

where I is equipped with only constant paths as dipaths, and the equivalence relations $\approx_t$ used to define the above quotient d-spaces are defined by $(s_1, t_1) \approx_t (s_2, t_2)$ if and only if $t_1 = t = t_2$. The **universal dicovering** of a pointed d-space (X, x)

is the pointed dicovering $p : (\tilde{X}, \tilde{x}) \to (X, x)$ such that for every dicovering $q :$ $(Y, y) \to (X, x)$ there exists a unique pointed morphism $r : (\tilde{X}, \tilde{x}) \to (Y, y)$ satisfying $p = q \circ r$.

In the above definition, three kinds of unique lifting properties are required: the first amounts to impose that paths lift uniquely once given their initial point, the last that endpoint-preserving homotopies between paths lift uniquely once given the initial point. The second lifting property is more subtle, and required in order to obtain lifting of more than just directed paths and dihomotopies, see Proposition 4.76. We shall first explain a bit more its effect. The space $(I \times \vec{I})/\approx_0$ is isomorphic to the "quarter disk" Q (also called a "fan") which is the subset of $\mathbb{C}$ consisting of points $re^{i\theta}$ such that $0 \leqslant r \leqslant 1$ and $0 \leqslant \theta \leqslant \pi/2$, and directed paths are those of the form $t \mapsto r(t)e^{i\theta}$ with $r : I \to I$ increasing and $\theta \in [0, \frac{\pi}{2}]$ (not depending on t). The universal dicovering of Q is the identity map. If we remove the second lifting property, the universal dicovering would be the space $\coprod_{\theta \in [0, \frac{\pi}{2}]} \vec{I}/\approx_0$, which has the same points and directed paths as Q but is equipped with a much finer topology.

Remark 4.67 These conditions are enough to imply that the universal dicovering is also the identity for higher dimensional "fans," e.g. , if Q is the positive octant of a 3-disk with a similar d-structure, see Proposition 4.76.

Remark 4.68 If the pointed d-space (X, x) is *well-pointed*, i.e., every point in X is the target of a dipath whose source is the base point, then instead of requiring unique lifting of dipaths in (X, x), we could equivalently only require unique lifting of dipaths with the base point x as source.

Example 4.69 Consider the d-space X obtained as the geometric semantics of a program of the form A; `while` b `do` B. As described in Example 4.17, the space X consists of a directed circle S^1 glued to a directed interval $[0, 2]$ by identifying $1 \in S^1$ to $1 \in [0, 2]$, as depicted on the right below. On the left is shown the associated universal dicovering $\tilde{X}$:

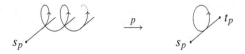

It can be shown that it corresponds to the geometric semantics of the infinite program obtained as the syntactic unfolding of the program p, as explained in the beginning of the section.

Example 4.70 Consider the geometric semantics X of the program $p = \mathsf{P}_a ; \mathsf{V}_a \parallel \mathsf{P}_a ; \mathsf{V}_a$ where a is a mutex (see Example 4.15). We have $X = \vec{I} \times \vec{I} \setminus [\frac{1}{4}, \frac{3}{4}]^2$. Its universal dicovering has two copies of the upper right square:

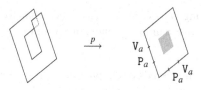

This example shows that in the universal dicovering, some nondirected loops are also unfolded.

Example 4.71 Consider the "directed box without bottom" B, which is the subspace of $\vec{I}^3$ consisting of points (x_1, x_2, x_3) such that $x_1 \in \{0, 1\}$ or $x_2 \in \{0, 1\}$ or $x_3 \in \{1\}$, with $x = (0, 0, 0)$ as base point, as shown on the right:

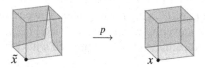

We have depicted the universal dicovering p. Note that it is the identity when restricted to each of the five squares forming the boundary of the box. The fiber $p^{-1}(x_1, x_2, x_3)$ always has one element, except when $x_1 = x_2 = 1$ and $0 \leqslant x_3 < 1$, in particular p is not a covering, if we consider B as a topological space. Also note that the underlying topological space is homeomorphic to a disk, and therefore the universal covering in the classical sense is the identity map: this illustrates the fact that distinct morphisms of the directed fundamental category do not always come from distinct morphisms of the fundamental groupoid.

Example 4.72 Consider the directed disk $\vec{D}^2$, as described in Example 4.4, together with 0 as basepoint. The associated universal dicovering is the identity map id : $\vec{D}^2 \to \vec{D}^2$. The universal dicovering space thus contains nontrivial directed loops, such as the path $g : 1 \to 1$ such that $g(t) = e^{2i\pi t}$. However, all the loops reachable from the basepoint are trivial: for instance, if we consider the path $f : 0 \to 1$ defined by $f(t) = t$, it can easily be shown that the path $f . g$, obtained as the concatenation of f and g is dihomotopic to f.

The universal dicovering of a pointed space always exists for abstract reasons [42, 53]. Moreover, there is an explicit construction along the lines of Proposition 4.64. However, the topology on the universal dicovering is constructed to satisfy a topological condition defined as follows:

Definition 4.73 A space (or a d-space) X is Δ-*generated*, if the topology is such that $U \subseteq X$ open if and only if $f^{-1}(U)$ is open for every map $f : \Delta^n \to X$.

Remark 4.74 A space X is Δ-generated if and only if it is *I-generated*, i.e., a subset $U \subseteq X$ is open if and only if $f^{-1}(U)$ is open for every path $f : I \to X$, see [45], which in turn is equivalent to requiring that a map $g : X \to Y$ is continuous if and only if $g \circ f$ is continuous for every map $f : I \to X$.

Any pointed d-space always admits a universal dicovering which can be constructed using a variant of the construction provided in Proposition 4.64.

Proposition 4.75 *Given a pointed d-space* (X, x), *its universal dicovering space is isomorphic to the pointed space* $(\tilde{X}, \tilde{x})$ *whose points are classes* $[f]$ *of dipaths* $f : x \rightarrowtail y$ *originating in* x *under endpoint-preserving dihomotopies, with* $\tilde{x}$ *the constant path* $\tilde{x} : x \rightarrowtail x$, *and the dicovering* $p : (\tilde{X}, \tilde{x}) \to (X, x)$ *associating to each class of a dipath* $f : x \rightarrowtail y$ *its endpoint* y. *The topology on* $\tilde{X}$ *is the* Δ-*generated topology generated by sets* $U_{[f]}$, *where* $U \subseteq X$ *is an open set and* $[f] \in \tilde{X}$ *is a dihomotopy class of paths in* X *such that* $f(1) \in U$, *defined by* $U_{[f]} = \left\{ [g] \in \tilde{X} \mid f \sim_U g \right\}$. *Here,* $f \sim_U g$ *means that* f *can be continuously deformed into* g *through dipaths whose source is* x *and target lies within* U.

When the base space X is Δ-generated, it can moreover be shown that the universal dicovering has the following unique lifting property:

Proposition 4.76 *Suppose given a universal dicovering* $p : (Y, y) \to (X, x)$, *with* X Δ-*generated, and a pointed* Δ-*generated d-space* (Z, z). *If for every* $z' \in Z$ *there is exactly one path up to dihomotopy from* z *to* z', *and if for every path* $f : I \to Z$ *there is a "fan" map* $h : (I \times \vec{I})/\approx_0 \to Z$ *such that* $h(0, 0) = z$ *and* $h(-, 1) = f$, *then every pointed map* $g : (Z, z) \to (X, x)$ *lifts uniquely to* (Y, y).

Proof The lifting $\hat{g} : (Z, z) \to (Y, y)$ of g is given by the unique lifting of all dipaths with source z: the point $\hat{g}(z')$ is the endpoint of the unique lifting of a dipath from z to z'. The endpoint of the lifting does not depend of the choice of dipath, since there is only one up to dihomotopy and dihomotopies with fixed endpoints lift. Since Z is Δ-generated, to show continuity of the lift, it suffices to see that $\hat{g} \circ f$ is continuous for every path $f : I \to Z$. Since there is a fan $h : (I \times \vec{I})/\approx_0 \to Z$ such that $h(0, 0) = z$ and $h(-, 1) = f$ and since fans lift uniquely (and continuously), $\hat{h}$ is the restriction of $\hat{g}$ and thus $\hat{g} \circ f$ is continuous. $\qquad\square$

Example 4.77 Let $f : I \to \vec{I}^n$ be a path, then $h(t, s) = sf(t)$ is a fan as above. The topology on the n-cube is clearly Δ-generated, so $(\vec{I}^n, 0)$ satisfies the condition above.

Example 4.78 (*Necklace of dicubes*) Consider the space X obtained from k dicubes $\vec{I}^{n_i}$, with $i \in [1 : k]$, by identifying $1 \in \vec{I}^{n_i}$ with $0 \in \vec{I}^{n_{i+1}}$, with basepoint $0 \in \vec{I}^{n_1}$. Also, consider a path $f : I \to X$. The fan is given as follows. Suppose $f(I) \subseteq \vec{I}^{n_j}$, then $h(t, s)$ is a concatenation of a dipath g from the basepoint to $0 \in \vec{I}^{n_j}$, $h(t, s) = g(2t, s)$ for $0 \leqslant t \leqslant \frac{1}{2}$ with a fan, as in Example 4.77, in $\vec{I}^{n_j}$. Since the topology on the necklace is given by the topology on each dicube plus the gluing, and the gluing is preserved in the lift, it suffices to study paths in one dicube.

Remark 4.79 The lifting properties we get here are similar to the classical case. For a covering $p : (Y, y) \to (X, x)$, a map $f : (Z, z) \to (X, x)$ where Z is path-connected and locally path-connected, lifts to a map $\hat{f} : (Z, z) \to (Y, y)$ if and only if the image of the fundamental group $f_*(\pi_1(Z, z))$ is a subgroup of $p_*(\pi_1(Y, y))$. In particular, if p is the universal covering, $p_*(\pi_1(Y, y))$ is trivial, so f lifts if and only if every loop in Z is contractible in the image [84]. For example, consider the fourfold covering of the circle Example 4.62 $p : (S^1, 1) \to (S^1, 1)$ given by $p(e^{2i\pi t}) = e^{8i\pi t}$ and let $f : (S^1, 1) \to (S^1, 1)$ be given by $f(e^{2i\pi t}) = e^{16i\pi t}$. This lifts to the map $\hat{f}(e^{2i\pi t}) = e^{4i\pi t}$, looping the covering space twice. The map $g(e^{2i\pi t}) = e^{6i\pi t}$ does not lift, since 4 does not divide 3. Similarly in dicoverings with less unfolding, more maps lift. Moreover, if (Z, z) is Δ-generated, if for every path $f : \vec{I} \to Z$ there is a "fan" map $h : (I \times \vec{I})/\approx_0 \to Z$ such that $h(0, 0) = z$ and $h(-, 1) = f$, if $p : (Y, y) \to (X, x)$ is a dicovering and $g : (Z, z) \to (X, x)$ is a pointed d-map such that $g_*(\vec{\Pi}_1(Z)(z, w))$ has only one element for all $w \in Z$, then g lifts. To see this, for $z' \in Z$, choose a dipath f from z to z'. The dipath $g \circ f$ lifts uniquely to a dipath $\widehat{g \circ f}$ and $\hat{g}(z) = \widehat{g \circ f}(1)$. As above, this is unique, since the dipath $g \circ f$ is unique up to dihomotopy. To see that the lift is continuous, use the lift of fans as above. In fact, the requirement on fans may be relaxed as follows. For every path $f : I \to Z$, there is a map $h : (I \times \vec{I})/\approx_0 \to Z$ such that $h(0, 0) = z$ and $h(-, 1) = f$ and $h(s_0, t)$ are dipaths in Z, h is not necessarily continuous, but $g \circ h$ is a d-map. The fan $g \circ h$ then lifts uniquely and again we may conclude that $\hat{g}$ is continuous.

Finally, from Proposition 4.75, it can be observed that the notion of universal dicovering is an extension of the classical notion of universal covering for a large class of spaces.

Proposition 4.80 *Suppose given a well-connected and Δ-generated space (X, x) with $(\tilde{X}, \tilde{x})$ as universal covering. If we see the space X as a d-space equipped with $dX = X^I$, i.e., if every path is directed, then $\tilde{X}$ is the underlying space of its universal dicovering.*

Remark 4.81 Note however that for an arbitrary d-space, the underlying space of one of its dicoverings is not generally a covering, as illustrated by various examples above.

Remark 4.82 The definition of universal dicovering is certainly not the most immediate generalization of the notion of universal covering one could think of, but other generalizations lack some nice properties. For instance, one could consider an immediate directed variant of covering spaces, as introduced in Definition 4.55, and define a *d-covering* map (we use this terminology in this remark only, to distinguish those from the dicoverings of Definition 4.66) as a morphism $p : Y \to X$ of d-spaces such that every point of X has a neighborhood U whose preimage is a disjoint union of open sets dihomeomorphic to U though p. Applying this definition to the examples mentioned above shows some differences with dicoverings, as defined in this section. First, this construction adds unnecessary "pasts" for every point: for instance, consider the directed square $\vec{I}^2 \setminus]0, 1[^2$, which is typically obtained as the geometric semantics of a program of the form `if` b `then` A `else` B (this is a variant of Example 4.70), its universal d-covering is

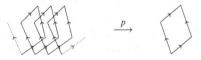

Even when restricting to points which are reachable from the base point, the resulting space is not satisfactory. In the "box without bottom" Example 4.71, the identity map is the only d-covering (and thus the universal one). This illustrates the fact that we get really different results with the two definitions (even if we restrict to the reachable part), and thus the construction given by Proposition 4.75, which is very useful in practice, cannot be used to construct the universal d-covering. Moreover, in this example, if we write y for the point $(1, 1, 0)$, there are two non-dihomotopic dipaths from x to y, which shows that the d-covering approach does not really "resolve" the part of the fundamental category consisting of paths starting from x. Another illustration of the difference of our construction and the d-covering is the disk with a directed boundary circle. Consider the unit disk $X = \{z \in \mathbb{C} \mid |z| \leqslant 1\}$ such that "only the boundary is directed," the directed paths in dX are either constant paths or paths of the form $t \mapsto e^{i\theta(t)}$ where θ is increasing modulo 2π. The boundary loop is not contractible, since there are no dipaths in the interior of the disk. The associated universal d-covering $\tilde{X}$ is the space itself (together with the identity map). In particular the directed boundary loop is not unrolled. The universal dicovering with basepoint $1 \in \mathbb{C}$ is the directed half-line $\mathbb{R}_+$ with covering map $p(t) = e^{i2\pi t}$, where the directed loop is unrolled.

4.4 Historical Notes and Other Models

For some time, this geometric view on processes has been scarcely used, apart from deriving some algorithms for finding deadlocks [24, 28] and determining serializability of transaction systems [118]. It came back to light due to considerations on truly concurrent semantics [141], and the links with the cubical models presented in Chap. 3.

The book by Grandis on d-spaces [77] is the first (and up to now the only) mathematical book on directed algebraic topology. His approach based on d-spaces has emerged as the most general and tractable one, however many other ways of formalizing the notion of direction have been explored and we briefly mention some of them here. Historically, the notion of *partially ordered space* (or *pospace*) came first [37, 128], see Example 4.3, originating in the study of positive cones of topological vector spaces occurring in functional analysis. Those spaces are convenient to work with, but their main drawback is that they cannot represent spaces containing nontrivial directed loops, see Example 4.4. In order to overcome this limitation, the notion of *local pospace* was introduced by some of the authors [51], which is a suitable sheaf of pospaces, i.e., a space such that every point admits a neighborhood with a pospace structure in a coherent way; for a gentle introduction to sheaves, see [111]. Unfortunately, the resulting category is rather ill-behaved: it is finitely complete but lacks infinite products (a product of local pospaces exists iff all but finitely many terms are actually pospaces); the natural embedding of the category of pospaces into the category of local pospaces is not full; some colimits of local pospaces do not preserve the topology (this is due to the fact that a local pospace has no vortex, see Example 4.4). Nevertheless most finite precubical sets can be realized in the category of local pospaces [51]: this category is nice enough if we restrict to computer-scientific applications. In order to encompass directed spaces with vortexes, the notion of *stream* was introduced by Krishnan [105]. This is a cosheaf of locally preordered spaces: a stream consists of a topological space such that each open subset U is equipped with a preorder in such a way that for all open coverings $(U_i)_{i \in I}$ of U, the preorder on U is the least preorder containing all the preorders on the U_i. This model turns out to be very close to d-spaces presented in this chapter: there is an adjunction between the categories of streams and d-spaces, which induces an isomorphism between subcategories that can be characterized explicitly [87]. The preceding list is by no means exhaustive: approaches based on model categories [21, 57–59, 168], and locally presentable categories [53] have also been investigated in this context.

The link between "continuous" dihomotopies in topological models and "discrete" deformations in combinatorial models are reminiscent of (simplicial) approximation theorems, in the classical case. Indeed, Theorem 4.38 is a particular case of a general cubical approximation theorem [44, 106, 171]. Note that Proposition 4.40 has many interesting consequences, and is in particular instrumental in the work on directed homology by Dubut and coauthors [35]. The directed Seifert–van Kampen theorem, Theorem 4.52 was first introduced for dihomotopy in a very restricted case [66], and later proved in the general case for d-homotopy [76]. The theory about dicoverings and the existence of a universal dicovering, see Proposition 4.75, was developed by Fajstrup in [45] and later reformulated by Krishnan and coauthors [69]. Such generalizations of coverings based on lifting properties have also been investigated in the nondirected setting [18]. As a matter of fact, different choices can be made. For instance, a theory of coverings has been developed for general streams (with also particular applications to streams that are geometric realizations of precubical sets), which are coverings in the usual sense of the underlying topological spaces [69]. This is not the case in general for the constructions in this chapter, but there is still a suitable lifting property for dipaths modulo dihomotopy in both cases, making them useful for applications. We should also mention that "unfolding" constructions, very similar to dicoverings, are classical in models for concurrency such as Petri nets and event structures [62, 132, 167].

Chapter 5
Algorithmics on Directed Spaces

In this section, we explain algorithms which are based on the geometric semantics of programs. In order to ease the presentation, those algorithms are formulated for simple programs, as defined in Sect. 4.1.4, and we only hint at generalizations: programs with branchings can generally be handled by adapting the algorithms, and loops can be handled up to a finite depth by unrolling the programs (see Remark 2.26 and Sect. 4.3.2). We have chosen to illustrate the wide variety of applications of the geometric point of view by presenting a compact way of representing regions in the geometric semantics (Sect. 5.1), an algorithm for detecting deadlocks (Sect. 5.2), and an algorithm for factoring programs into independent parallel processes (Sect. 5.3). These will also be used in subsequent chapters in order to compute components (Sect. 6.3) and path spaces (Sect. 7.1.4). Most of the techniques described here have been implemented in the tool ALCOOL developed by some of the authors of the book [67]. Directed topological models have found applications to validation and static analysis [12, 48, 67, 71], serializability and correctness of databases [82], and fault-tolerant protocols for distributed systems [75]. A panorama of applications can be found in [51, 66] and in Chap. 8.

Throughout the section, we will consistently use the notations introduced in Sect. 4.1.4 for simple programs: given such a program p of dimension n, its geometric semantics $\check{G}_p$ which we denote X in the following, is of the form

$$X = \check{G}_p = \vec{I}^n \setminus \bigcup_{i=1}^{l} R^i \quad \text{with} \quad R^i = \prod_{j=1}^{n}]x_j^i, y_j^i[\tag{5.1}$$

with, for every $i \in [1 : l]$ and $j \in [1 : n]$, $x_j^i, y_j^i \in \{-\infty\} \cup I \cup \{\infty\}$ and $x_j^i < y_j^i$.

© Springer International Publishing Switzerland 2016
L. Fajstrup et al., *Directed Algebraic Topology and Concurrency*,
DOI 10.1007/978-3-319-15398-8_5

5.1 The Boolean Algebra of Cubical Regions

We consider here simple programs of dimension n, as mentioned in the introduction of this chapter. In this situation, the space X is thus a cubical region in the following sense:

Definition 5.1 Given a space $X \subseteq \vec{I}^n$, a *cubical cover* $R = (R^i)_{1 \leq i \leq l}$ of X is a finite family of n-cubes (with open or closed boundaries) in X such that $\bigcup R = X$. A space which admits a cubical cover is called a **cubical region**. We write $\mathscr{C}_n$ (resp. $\mathscr{R}_n$) for the set of cubical covers (resp. regions) of dimension n.

Note that in this section, R will refer to a such a cubical cover, and not only to the cover of the forbidden region (which is an instance of a cubical region).

Remark 5.2 We insist on the fact that we are considering here cubes which can have both open and closed boundaries, which makes some properties a bit more difficult to formulate than if we had assumed that they were all open (or closed). For instance, given two cubes, their union may be path-connected even if they have an empty intersection. Moreover, the intervals defining the cubes may not have equal length.

Since cubical covers are supposed to be finite, they provide a way to represent the associated cubical regions and manipulate them algorithmically. In this section, we investigate the operations available on such representations.

Proposition 5.3 *The set of cubical regions is closed under union, intersection, and complement in $\vec{I}^n$. It is thus a Boolean subalgebra of the powerset $\mathfrak{P}(\vec{I}^n)$.*

Proof Given two cubical covers R and S of $X = \bigcup R$ and $Y = \bigcup S$, respectively, the spaces $X \cup Y$ and $X \cap Y$ are both cubical regions since they, respectively, admit $R \cup S$ and $\{R^i \cap S^j \mid R^i \in R$ and $S^j \in S\}$ as covers, the later definition relying on the fact that the intersection of two n-cubes is an n-cube and that intersection distributes over union. Finally, the complement of an n-cube C is easily shown to admit a cubical cover C^c (see below) from which follows that the complement X^c of X can be covered by $\bigcap \{(R^i)^c \mid R^i \in R\}$ (the intersection is in the sense of cubical regions defined above). $\qquad\square$

A cubical region generally admits multiple cubical covers, however there is always a canonical one which represents the cubical region, and usual operations can be performed quite efficiently on this representation. Given two cubical covers R and S of the same cubical region X, we write $R \preceq S$ whenever for every n-cube $R^i \in R$ there exists an n-cube $S^j \in S$ such that $R^i \subseteq S^j$. It can be shown that the poset of cubical covers of X admits a maximum element, called the *normal form* of the cubical region: it consists of the maximal n-cubes included in the region. We write $\overline{\mathscr{C}}_n \subseteq \mathscr{C}_n$ for the set of cubical covers in normal form.

Example 5.4 Consider the region shown on the left below:

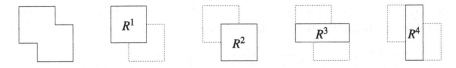

Its cubical cover in normal form is $R = \{R^1, R^2, R^3, R^3\}$. Note that it is not a cubical cover with fewest elements, since $\{R^1, R^2\}$ is also a cover for instance. It is however "quite small" since adding to R any rectangles included in the region will yield another cubical cover.

The situation can be presented more conceptually as follows:

Proposition 5.5 *The sets $\mathscr{C}_n$ (resp. $\mathscr{R}_n$) can be seen as posets when equipped with the partial order $\preceq$ (resp. inclusion). The functions $U_n : \mathscr{C}_n \to \mathscr{R}_n$ such that $U_n(R) = \bigcup R$, and $M_n : \mathscr{R}_n \to \mathscr{C}_n$ associating to X the set $M_n(X)$ of maximal n-cubes in X, form a Galois connection (with U_n left adjoint to M_n).*

Since it can easily be checked that we have $U_n \circ M_n = \mathrm{id}_{\mathscr{R}_n}$, we deduce that

Proposition 5.6 *The Galois connection of Proposition 5.5 induces a bijection between $\mathscr{R}_n$ and the subposet $\overline{\mathscr{C}}_n$ of $\mathscr{C}_n$ whose elements are covers in normal form.*

Given an n-cube $C = \prod_{j=1}^{n}]x_j, y_j[$, the normal form of its complement is given by
$$C^{\mathrm{cmax}} = \{ \ldots \times I \times [0, x_j] \times I \times \ldots, \ldots \times I \times [y_j, 1] \times I \times \ldots \mid j \in [1:n]\}$$
and similar formulas can be given when some of the boundaries are closed. We say that a cover is *prenormal* if it contains the associated normal cover: such a cover can easily be converted into a normal one by removing cubes included in others. It can be shown that the intersection and the complement of a prenormal cover, as described in the proof of Proposition 5.3, are still prenormal (if we use the above complement for individual cubes when computing the complement). Since we have the normal form of the complement of any cube, a cover can be turned into a prenormal one by complementing it twice, and the union of two prenormal covers R and S can be computed as $(R^c \cap S^c)^c$ in order to preserve prenormality. In the following, we will also use the operation $R \setminus S = R \cap S^c$.

When the geometric semantics of a program is a cubical region $X = \bigcup R$ described by a normal cubical cover R, its unsafe and doomed regions (see Definition 4.44) can be computed as in Algorithm 5.7. We first need some definitions. We introduce a partial order $\lhd$ on the elements R^i of R as the reflexive and transitive closure of the relation such that $R^i \lhd R^{i'}$ iff $R^i \cup R^{i'}$ is connected and $R^{i'}$ contains a point which is strictly above every point of R^i w.r.t. to the product order. In this case, we say that R^i is *in the past* of $R^{i'}$: this means that there are some points of $R^{i'}$ which are not in R^i but reachable from any point in R^i (in the case where R^i and $R^{i'}$ are both open, or both closed, we have $R^i \lhd R^{i'}$ iff $x_j^{i'} \in R^i$ and $y_j^i \in R^{i'}$). Above, by "$R^i \cup R^{i'}$ is connected," we mean here that there is a nondirected path from every point of R^i to every point of $R^{i'}$: this can be algorithmically decided by suitably comparing the boundaries, taking care of whether they are open or not. Given a region R, we write $\triangle R$ for its downward closure w.r.t. $\lhd$. The unsafe and doomed regions of a simple program can then be computed as follows:

Algorithm 5.7 Given the normal cubical cover R of X:

1. compute the cover $\mathcal{U}(R)$ which consists of cubes $R^i \in R$ such that

    – R^i does not contain the maximal point of $\vec{I}^n$, and
    – R^i is maximal w.r.t. $\lhd$,

2. a cover of the unsafe region can then be obtained as $\triangle \mathcal{U}(R)$,
3. a cover of the doomed region is $\mathcal{D}(R) = \triangle \mathcal{U}(R) \backslash \triangle \mathcal{E}(R)$ where $\mathcal{E}(R)$ is the set of rectangles in R containing the maximal point of $\vec{I}^n$.

Example 5.8 The geometric semantics of the Swiss flag program described in Example 4.21 contains eight maximal cubes. Two of them are shown on the left and middle picture below, and the others can be obtained by symmetry.

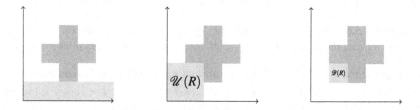

Notice that $\mathcal{U}(R)$ is reduced to the square R^1 displayed in the middle picture and the doomed region $\mathcal{D}(R)$ is therefore reduced to the square on the right. The region $\triangle \mathcal{E}(R)$ consists of all the points except the dead ones (see figure on the right).

We have only briefly presented cubical regions as subspaces of $\vec{I}^n$, but this can easily be generalized to situations such as $|G_1| \times \ldots \times |G_n|$, where all the G_i are finite graphs, thus allowing the handling of programs with loops: this stems from the remark that if G is a finite graph, then the collection of finite unions of connected subsets of its geometric realization $|G|$ forms a Boolean subalgebra of the powerset Boolean algebra $\mathfrak{P}(|G|)$. The case developed in this section is the particular case where all the graphs G_i are isomorphic to the graph with two vertices and one edge between them.

Example 5.9 The "Swiss torus" is the space obtained by removing a cross as in the Swiss flag example, but on a torus instead of $\vec{I}^2$. It can be modeled as the product of the following graphs G_1 and G_2:

$$G_1 \quad = \quad \begin{array}{c} P_b \xrightarrow{x_{ab}} V_b \\ x_a \uparrow \qquad \downarrow x'_a \\ P_a \xleftarrow{\ \ } V_a \\ x \end{array} \qquad\qquad G_2 \quad = \quad \begin{array}{c} P_a \xrightarrow{y_{ab}} V_a \\ y_a \uparrow \qquad \downarrow y'_a \\ P_b \xleftarrow{\ \ } V_b \\ y \end{array}$$

The space $|G_1| \times |G_2|$, along with the forbidden region, is drawn on the left below: in this representation, the parallel (vertical or horizontal) external edges should be identified.

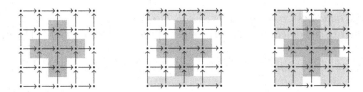

The maximal cover of the resulting space has three "rectangles:" one is shown in the middle (notice that the region is in fact connected since a point on the boundary below should be identified with the corresponding point on the boundary above), one is obtained from the previous one by a rotation, and the last one is shown on the right.

Efficient algorithms for representing cubical regions, and computing intersections of cubes in particular, have appeared in numerous contexts such as in computational geometry [142]. Unions of isothetic hypercubes as we use them here are called "orthogonal polyhedra" in computational geometry and have also been used in hybrid systems theory [16].

5.2 Computing Deadlocks

We provide an algorithm to detect deadlocks in simple programs, and more generally doomed regions, which is based on the geometrical characterization of deadlocks, in the sense of Definition 4.44. This simple algorithm, introduced in [50], is more efficient than the one based on cubical regions given in the previous chapter, and will be the basis for the algorithms computing the trace space of programs in Chap. 7.

As in the previous section, we are interested in a simple program p of dimension n, whose geometric semantics is of the form (5.1). In order to be as general as possible, we generally do not make further hypothesis on the spaces we consider. However, we will sometimes need the following property, which can easily be shown to be satisfied for the geometric semantics of simple programs:

Definition 5.10 A d-space X of the form (5.1) is *generic* when given two distinct indices $i, i' \in [1 : l]$, if x_j^i and $x_j^{i'}$ are both equal, and different from $-\infty$ then $R^i \cap R^{i'} = \emptyset$.

The careful reader will have noticed that, to be precise, the genericity condition applies not to X, but to the given cubical cover R of its complement. The main result of this section is the following characterization of deadlock points.

Theorem 5.11 *Suppose that* $y_j^i \neq \infty$ *for every* $i \in [1 : l]$ *and* $j \in [1 : n]$. *Given a point* $z \in X$, *such that* $z \notin \bigcup_{i=1}^{l} R^i$, *the following are equivalent:*

(i) *z is a deadlock*
(ii) *there exists a function* $i : [1 : n] \to [1 : l]$ *such that for every* $j \in [1 : n]$,

$$z_j = x_j^{i(j)} \quad and \quad \forall j' \in [1 : n], \ j' \neq j \ \Rightarrow \ x_j^{i(j')} < x_j^{i(j)} < y_j^{i(j')}$$

If the space is supposed to be generic, this is moreover equivalent to

(iii) *there exists an injective function* $i : [1 : n] \to [1 : l]$ *such that for every* $j \in [1 : n]$,

$$z_j = x_j^{i(j)} \quad and \quad \forall j' \in [1 : n], \ x_j^{i(j')} \leq x_j^{i(j)} < y_j^{i(j')}$$

(iv) *there exists a subset* $L \subseteq [1 : l]$ *of cardinal n satisfying*

$$\bigcap_{i \in L} R^i \neq \emptyset \quad and \quad z = \inf \left(\bigcap_{i \in L} R^i \right)$$

and in this case $z_j = \max \left\{ x_j^i \mid i \in L \right\}$ *for every* $j \in [1 : n]$.

Proof We show the required implications.

(i) $\Rightarrow$ (ii) Suppose that z is a deadlock. Given $j \in [1 : n]$, if we write e_j for the unit vector of I^n in direction j, we have that $z + te_j$ belongs to some rectangle R^i for $t > 0$ small enough, and choosing such an index $i(j)$ for each direction j provides a suitable function $i : [1 : n] \to [1 : l]$ of indices of rectangles.

(ii) $\Rightarrow$ (i) Suppose given a suitable function $i : [1 : n] \to [1 : l]$, and consider a directed path f starting from z. We are going to show that this path is constant. Suppose that there exists $t \in I$ such that $f(t) \neq z$. There exists a direction $j \in [1 : n]$ such that $f(t)_j \neq z_j$, i.e. $z_j = x_j^{i(j)} < f(t)_j$ because the path f is directed. Moreover, for t small enough, we also have $f(t)_j < y_j^{i(j)}$ by continuity of f. Finally, for $j' \neq j$, we have $z_{j'} = x_{j'}^{i(j')}$, and therefore $x_{j'}^{i(j)} < z_{j'} \leq f(t)_{j'} < y_{j'}^{i(j)}$ for t small enough. We deduce that $f(t) \in R^{i(j)}$ which is absurd. The path f is thus constant and z is a deadlock.

(ii) $\Leftrightarrow$ (iii) Straightforward.

(iii) $\Rightarrow$ (iv) Suppose given a suitable function $i : [1 : n] \to [1 : l]$. We define L to be the image of i. Given $i \in L$ and $j \in [1 : n]$, we have $x_j^i \le z_j < y_j^i$. Therefore, for $t > 0$ small enough $z + \sum_{i=1}^{n} t e_i$ belongs to every R^i for $i \in L$ and $\bigcap_{i \in L} R^i \ne \emptyset$. The other conditions are easy to check.

(iv) $\Rightarrow$ (iii) Suppose given a suitable set $L \subseteq [1 : l]$. By hypothesis, we have
$$z_j = \inf \left(\bigcap_{i \in L} R_j^i \right) = \inf \left(\bigcap_{i \in L}]x_j^i, y_j^i[\right) \text{ and therefore } z_j = \max \left\{ x_j^i \mid i \in L \right\}$$
because $\bigcap_{i \in L} R^i \ne \emptyset$. Because of the genericity condition, there is exactly one index $i \in L$ such that $z_j = x_j^i$, that we denote by $i(j)$. The function $i : [1 : n] \to L$ thus defined is injective. Namely, given $i' \in L$ which is not in the image of i, we have, for every $j \in [1 : n]$, $x_j^{i'} < z_j < y_j^{i'}$ by definition of z, and therefore $z \in R^{i'}$, which contradicts the last hypothesis. Using a similar reasoning, one shows that, for every $j' \in [1 : n]$, we have $x_j^{i(j')} \le x_j^{i(j)} < y_j^{i(j')}$. $\qquad\square$

Remark 5.12 In the case (ii), the function i is not assumed to be injective, but one can actually deduce that it is always injective from the conditions imposed on it.

Some examples illustrating the above theorem are provided below. We should first notice that it allows us to directly formulate the following algorithm for detecting deadlocks:

Algorithm 5.13 When the d-space X is generic, the deadlock points can be found as follows:

1. find n intervals $R^{i_1}, \dots, R^{i_n}$ such that $\bigcap_{j=1}^{n} R^{i_j} \ne \emptyset$,
2. compute z defined by $z_j = \max \left\{ x_j^{i_1}, \dots, x_j^{i_n} \right\}$,
3. if for every $i \in [1 : l]$, $z \notin R^i$ then z is a deadlock.

There are of course many possible optimizations to this algorithm in order to avoid computing intersections multiple times, etc. These will not be detailed here, but the reader can find an optimized algorithm for the closely related algorithm for computing trace spaces in Chap. 7.

The preceding algorithm can be extended in order to compute the doomed region of a space. With the notations of the previous theorem, given a deadlock point z, the interval
$$]z', z] \subseteq \vec{I}^n \quad \text{with} \quad z_j' = \max \left\{ x_j^i \mid i \in I, x_j^i \ne z_j \right\}$$

for $j \in [1 : n]$, contains only doomed points, from which the deadlock z will eventually be reached. We call it the doomed interval associated to the deadlock z. By iterating the computation of such intervals, the doomed region of a simple program can be computed as follows:

Algorithm 5.14 The doomed regions of the space $X = \vec{I}^n \setminus \bigcup_{i=1}^{l} R^i$ can be found as follows:

1. find the deadlocks $z_1, \dots, z_m$ using Algorithm 5.13,
2. compute the associated doomed intervals $U_k =]z_k', z_k]$ for $k \in [1 : m]$ as above,

3. return those doomed intervals U_k as well as the doomed region of $X \setminus \bigcup_{k=1}^{m} U_k$ using the algorithm recursively (notice that the intervals U_k are not open on the right, however the previous algorithms can be straightforwardly adapted to handle those).

Example 5.15 Consider the program $p = P_a; P_b; V_a; P_c; V_c; V_b \parallel P_c; P_a; V_a; P_b; V_b; V_c$ whose geometric semantics $\check{G}_p$ is shown on the left:

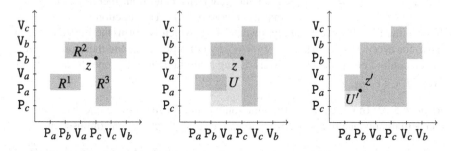

This geometric semantics is of the form $\vec{I}^2 \setminus \bigcup_{i=1}^{3} R^i$ with $R^1 =]\frac{1}{7}, \frac{3}{7}[\times]\frac{2}{7}, \frac{3}{7}[$, $R^2 =]\frac{2}{7}, \frac{6}{7}[\times]\frac{4}{7}, \frac{5}{7}[$ and $R^3 =]\frac{4}{7}, \frac{5}{7}[\times]\frac{1}{7}, \frac{6}{7}[$. The point $z = (\frac{4}{7}, \frac{4}{7})$ is a deadlock: one can apply Theorem 5.11 with $I = \{2, 3\}$, or $i(1) = 3$ and $i(2) = 2$. The doomed region U associated to z is also shown. Notice that adding the doomed region U to the forbidden region (as shown on the right) produces a new deadlock z' with associated forbidden region U'.

Remark 5.16 Consider the space $X = \vec{I}^2 \setminus \bigcup_{i=1}^{2} R^i$ with $R^1 =]\frac{1}{5}, \frac{4}{5}[\times]\frac{1}{4}, \frac{2}{4}[$ and $R^2 =]\frac{2}{5}, \frac{3}{5}[\times]\frac{1}{4}, \frac{3}{4}[$:

Notice that the point $z = (\frac{2}{5}, \frac{1}{4})$ satisfies the conditions of Theorem 5.11 (iii) or (iv) with $L = \{1, 2\}$, or $i(1) = 2$ and $i(2) = 1$. However, z is not a deadlock. In fact, the theorem does not apply because the space X is not generic: we have $x_2^1 = x_2^2$.

Example 5.17 Consider the following program, from Lipski and Papadimitriou [118], where a, b, c, d, e, and f are mutexes:

$$P_a; P_b; P_c; V_a; P_f; V_c; V_b; V_f \quad \parallel \quad P_d; P_e; P_a; V_d; P_c; V_e; V_a; V_c$$
$$\parallel \quad P_b; P_f; V_b; P_d; V_f; P_e; V_d; V_e$$

Its geometric semantics is represented below, together, on the right-hand side, with the request graph, a common tool for proving absence of deadlocks (see below).

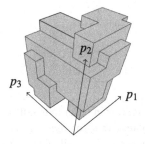 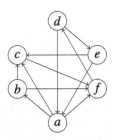

The *request graph* is constructed as follows: it has a node for every shared resource, and there is an edge from a to b when a process has acquired a lock on a without relinquishing it and is requesting a lock on b. For programs with mutexes, having an acyclic request graph implies deadlock freedom. We can see with this example that the converse is not true: the request graph for the Lipski and Papadimitriou program is cyclic, but there is no deadlock. The following path never deadlocks and goes all the way through the forbidden region:

$$P_b^3 ; P_f^3 ; V_b^3 ; P_a^1 ; P_d^2 ; P_e^2 ; P_b^1 ; P_c^1 ; V_a^1 ; P_a^2 ; V_d^2 ; P_d^3 ; V_f^3 ; P_f^1 ; V_c^1 ; V_b^1 ; V_f^1 ; P_c^2 ; V_e^2 ; V_a^2 ; V_c^2 ; P_e^3 ; V_d^3 ; V_e^3$$

In fact, the forbidden region is homotopically equivalent to the circle, and the dipath above goes through the interior of the circle, which is not obvious when first observing the program or its geometric semantics. Algorithm 5.13 indeed finds that all generic intersections of three forbidden cubes are included in other forbidden cubes, hence do not account for deadlocking situations. The methods of Sect. 6.3 and of Sect. 7.1.3 will allow us to describe the seven dihomotopy classes of total dipaths.

Remark 5.18 The previous example is also an example of a *nonserializable* program. Serializable programs are those for which every execution trace is equivalent to a serial one, i.e., one which corresponds to the execution of each process, entirely, in some order. This is a classical correctness criterion for concurrent databases, see [9]. Of course, this implies that there should be at most $n!$ classes of dipaths for serializable programs, so for three processes as we have here we should have six, and not seven, classes of dipaths for it to be serializable. Examples of serializable programs are 2PL ("two-phase locking") programs where all processes lock first all resources they need, and then, in a second phase, unlock them all. For instance, the Swiss flag Example 3.22 and the dining philosophers Example 4.18 are such programs. The first geometric proof of serializability of 2PL programs was given in [82], and later in [51] using concepts from directed algebraic topology.

Remark 5.19 In Theorem 5.11, the assumption that $y_j^i \neq \infty$ is necessary because a hole might cause a deadlock on a boundary. For instance, in $\vec{I}^2 \setminus (]\frac{1}{3}, \infty[\times]\frac{1}{3}, \frac{2}{3}[)$, the point $z = (1, \frac{1}{3})$ is a deadlock, as shown on the right:

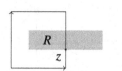 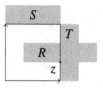

The theorem and the algorithm can easily be extended in order to handle those cases, by adding new holes "outside the maximal boundaries" of the space $\vec{I}^2$ (or $\vec{I}^n$ in general), as shown on the right (the regions S and T): the deadlock z is now detected with $I = \{R, T\}$, $i(1) = T$, and $i(2) = R$.

Generalizing the deadlock algorithm to programs with loops is straightforward: it is enough to unroll all loops once and find deadlocks in the resulting program. Finding the doomed region is not as easy: an overapproximation may be found by unrolling once, but to find the exact doomed region requires more than one unrolling in general as illustrated in following example:

Example 5.20 Consider the two following processes p and q, each containing a loop:

$$p = \mathrm{P}_d \,; \mathrm{P}_a \,; (\texttt{while } b \texttt{ do } p') \,; \mathrm{V}_a \,; \mathrm{P}_e \,; \mathrm{V}_d \,; \mathrm{V}_e$$
$$q = \mathrm{P}_e \,; \mathrm{P}_a \,; (\texttt{while } b' \texttt{ do } q') \,; \mathrm{V}_a \,; \mathrm{P}_d \,; \mathrm{V}_e \,; \mathrm{V}_d$$

with

$$p' = \mathrm{P}_b \,; \mathrm{V}_a \,; \mathrm{V}_d \,; \mathrm{P}_c \,; \mathrm{V}_b \,; \mathrm{P}_a \,; \mathrm{P}_d \,; \mathrm{V}_c \,; \mathrm{P}_b \,; \mathrm{V}_a \,; \mathrm{P}_c \,; \mathrm{V}_b \,; \mathrm{P}_a \,; \mathrm{V}_c \,; \mathrm{P}_b \,; \mathrm{V}_a \,; \mathrm{P}_c \,; \mathrm{V}_b \,; \mathrm{P}_a \,; \mathrm{V}_c$$
$$q' = \mathrm{P}_b \,; \mathrm{V}_a \,; \mathrm{P}_c \,; \mathrm{V}_b \,; \mathrm{P}_a \,; \mathrm{V}_c \,; \mathrm{P}_b \,; \mathrm{V}_a \,; \mathrm{P}_c \,; \mathrm{V}_b \,; \mathrm{P}_a \,; \mathrm{V}_c$$

The geometric semantics of $p \,\|\, q$ contains a torus (with holes) corresponding to the two loops in parallel, as shown on the left in the following pictures, which can be obtained by gluing parallel faces of the rectangle on the right:

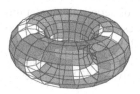 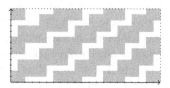

In a first unrolling, it seems that almost all states within the loops are doomed as shown on the left. However, on further unrolling, it becomes clear that no state in the loops is doomed. The dipaths from states in the loop to the final point may go via several iterations of each loop as illustrated on the right:

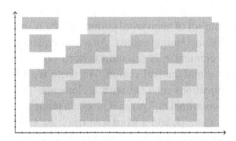 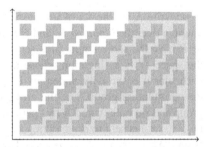

Geometrically, the above example is a torus knot, and similar examples may be constructed for higher dimensional knots. In such cases, the doomed region can be determined with a finite number of unrollings, and the criterion for computing the number of times the program should be unrolled can also be extended to the case of programs with nested loops [42].

5.3 Factorizing Programs

An important question, both from a theoretical and practical point of view, is whether a concurrent program can be decomposed as several processes which are running concurrently and are completely independent, i.e., the execution of one has no impact on the execution of the other one in parallel [34, 125]. We consider only simple programs for clarity.

Definition 5.21 In a program $p \parallel q$, the processes p and q are *independent* when $\check{C}_{p \parallel q} = \check{C}_p \otimes \check{C}_q$, or equivalently $\check{G}_{p \parallel q} = \check{G}_p \times \check{G}_q$.

In simple cases, independence can be detected syntactically. We write $\mathrm{FR}(p)$ for the set of resources used in a process p.

Lemma 5.22 *In a program $p \parallel q$ such that $\mathrm{FR}(p) \cap \mathrm{FR}(q) = \emptyset$, the processes p and q are independent.*

In the general case, this problem can be reduced to factorizing the geometric semantics of the program as a cartesian product of subspaces. An algorithm achieving this task is provided here. It has some similarities with the factorization of integers as products of primes. A more detailed presentation of this algorithm can be found in [6]. An important point to note is that we want to be able to factorize $p \parallel q \parallel r$ as $q \parallel (p \parallel r)$ for instance, when q and $p \parallel r$ are independent: this means that we have to consider the programs up to associativity, but also up to permutation of processes, i.e., up to the congruence $\approx$ such that $p \parallel q \approx q \parallel p$, etc., which is allowed because of Proposition 2.25.

Consider a simple program p of dimension n. We know from Sect. 5.1 that its geometric semantics $X = \check{G}_p$ is a cubical region and thus admits a cubical cover R.

If we write $\mathbb{A}$ for the set of subintervals of $\vec{I}$, each n-cube can be seen as a word of length n over the *alphabet* $\mathbb{A}$, and R as a finite set of such words. Since all the words of R have the same length, this set is said to be *homogeneous*. As a consequence, the space $\vec{I}^0$ is a singleton whose unique element ε is the empty word on $\mathbb{A}$. In particular it is the only nonempty 0-dimensional cubical region. We write $\mathfrak{S}_n$ for the permutation group, i.e., the group of bijections on a set with n elements. This group acts on both X and R by "permuting coordinates:" given $\sigma \in \mathfrak{S}_n$, a point $x = (x_1, \ldots, x_n)$ is sent to $\sigma \cdot x = (x_{\sigma^{-1}(1)}, \ldots, x_{\sigma^{-1}(n)})$, and a word $u = u_1 \ldots u_n \in \mathbb{A}^n$ is sent to $\sigma \cdot u = u_{\sigma^{-1}(1)} \ldots u_{\sigma^{-1}(n)}$. This actions obviously extends to cubical covers by $\sigma \cdot R = \{\sigma \cdot C \mid C \in R\}$.

We can define a (graded) tensor product on the symmetric groups as follows: given $\sigma \in \mathfrak{S}_p$ and $\tau \in \mathfrak{S}_q$, $\sigma \otimes \tau \in \mathfrak{S}_{p+q}$ is the permutation defined by

$$(\sigma \otimes \tau)(i) = \begin{cases} \sigma(i) & \text{if } 1 \le i \le p \\ \tau(i - p) + p & \text{if } p < i \le p+q \end{cases}$$

Since taking the cartesian product of two spaces amounts to concatenating the coordinates of their points (and similarly the product of two cubical regions amounts to concatenating the corresponding words), the group action is compatible with the cartesian product:

Lemma 5.23 *Given two spaces $X \subseteq \vec{I}^p$ and $Y \subseteq \vec{I}^q$, and two permutations $\sigma \in \mathfrak{S}_p$ and $\tau \in \mathfrak{S}_q$, we have $(\sigma \cdot X) \times (\tau \cdot Y) = (\sigma \otimes \tau) \cdot (X \times Y)$ (and similarly for regions).*

It thus makes sense to define the following monoids.

Definition 5.24 The *monoid of cubical regions* $\mathscr{R}^{\mathfrak{S}} = \coprod_{n \in \mathbb{N}} \mathscr{R}_n/\mathfrak{S}_n$ is the set of orbits of cubical regions under the action of $\mathfrak{S}_n$, equipped with the multiplication induced by cartesian product (i.e., concatenation), and its neutral element is the nonempty zero-dimensional cubical region. The *monoid of normal cubical covers* $\overline{\mathscr{C}}^{\mathfrak{S}} = \coprod_{n \in \mathbb{N}} \overline{\mathscr{C}}_n/\mathfrak{S}_n$ is defined similarly as the quotient of cubical covers in normal form under the action of the symmetric group.

These monoids are easily shown to be commutative. The isomorphisms between cubical regions and cubical covers in normal form of Proposition 5.6 induce a (graded) isomorphism between the above two monoids. We will thus only speak about the former in the following, but the algorithms are more naturally expressed when considering the latter.

Proposition 5.25 *The monoids $\mathscr{R}^{\mathfrak{S}}$ and $\overline{\mathscr{C}}^{\mathfrak{S}}$ are isomorphic.*

The monoid of cubical regions can be thought of as an analogue to the polynomial ring $\mathbb{k}[X_1, \ldots, X_n]$ over a ring $\mathbb{k}$. Now suppose that $\mathbb{k}$ is *factorial* (i.e., every element admits a factorization as a product of irreducible elements, which is unique up to reordering and multiplying by invertible elements), for instance $\mathbb{k} = \mathbb{Z}$. In this case, it is well known that the ring $\Bbbk[X_1, \ldots, X_n]$ is also factorial [112]. For instance, we have $\sum_{i=0}^{5} X^i = (X+1)(X^2 - X + 1)(X^2 + X + 1)$. However, this property fails for semirings: even though $\mathbb{N}$ is factorial, the ring $\mathbb{N}[X]$ is not, as illustrated by the following example from [130],

$$\sum_{i=0}^{5} X^i = (X^3 + 1)(X^2 + X + 1) = (X + 1)(X^4 + X^2 + 1)$$

because the factors P involved above, such as $P = X^3 + 1$, are *irreducible* (i.e., if $P = QR$ then either Q or R is invertible) but not *prime* (if P divides QR then P divides Q or P divides R, and P is not invertible). However, in the case of the monoid of regions this property holds (even though intuitively, it is closer to the case of polynomials with coefficients in $\mathbb{N}$): an n-cube (up to permutation) corresponds to a monomial of degree n, and a cubical region in $\mathscr{R}_n/\mathfrak{S}_n$ to a homogeneous polynomial of degree n.

Theorem 5.26 *In the commutative monoid $\mathscr{R}^{\mathfrak{S}}$ of cubical regions, every element can be uniquely factored as a product of irreducible elements.*

The decomposition of a cubical region can be performed algorithmically as follows. Given $u = u_1 \ldots u_n \in \mathbb{A}^n$ and a subset $I \subseteq [1 : n]$ of indices, we write $u|_I$ for the subword of u consisting of letters with indices in I. Given a homogeneous set of words $R \subseteq \mathbb{A}^n$, we also write $R|_I = \{u|_I \mid u \in R\}$.

Lemma 5.27 *Given $I \subseteq [1 : n]$, we write $I^c = [1 : n]\backslash I$. Given a homogeneous set of words $R \subseteq \mathbb{A}^n$, we have $R = R|_I \times R|_{I^c}$ (in the commutative monoid of cubical regions) if and only if for all words $u, v \in R$ there exists a word $w \in R$ such that $w|_I = u|_I$ and $w|_{I^c} = v|_{I^c}$.*

Whether a region can be factored can thus be tested by the following algorithm, which is a variant of the naive factorization algorithm for integers.

Algorithm 5.28 Given $R \subseteq \mathbb{A}^n$, a cubical cover of the state space in normal form:

1. choose a set $I \subseteq [1 : n]$ of cardinality $p \le n/2$,
2. compute the set $S = \{\pi_{I^c}(\pi_I^{-1}(u)) \mid u \in \pi_I(R)\} \subseteq \mathfrak{P}(\mathbb{A}^{n-p})$, where $\pi_I : R \to \mathbb{A}^p$ is the function such that $\pi_I(u) = u|_I$,
3. if S is a singleton then R factorizes as $R = R|_I \times R|_{I^c}$, otherwise try another set I.

Example 5.29 Suppose given resources a, b, c with $\kappa_a = \kappa_b = 1$ and $\kappa_c = 2$, and consider the program $p = p_1 \parallel p_2 \parallel p_3 \parallel p_4$ with

$$p_1 = \mathrm{P}_a\,;\mathrm{P}_c\,;\mathrm{V}_c\,;\mathrm{V}_a \quad p_2 = \mathrm{P}_b\,;\mathrm{P}_c\,;\mathrm{V}_c\,;\mathrm{V}_b \quad p_3 = p_1 \quad p_4 = p_2$$

The naive syntactic analysis of Lemma 5.22 would not discover that this program can be factored since all processes share the resource c. However, the following can be observed: thanks to the mutex a (resp. b), the processes p_1 and p_3 (resp. p_2 and p_4) cannot both hold the resource c at the same time. The resource c is thus never taken simultaneously more than two times, which means that the instructions P_c and V_c actually have no effect on the execution of the program. Moreover, if we remove the instructions P_c and V_c, the naive algorithm would detect that the resulting program can be factored in two independent processes using disjoint resources. The normal cubical cover R of the geometric semantics has 16 cubes (we are using $[0, 5]^4$ instead of $\vec{I}^4$ for readability):

$[0, 1[\times [0, 1[\times [0, 5] \times [0, 5]$ $[0, 1[\times [4, 5] \times [0, 5] \times [0, 5]$ $[0, 1[\times [0, 5] \times [0, 5] \times [0, 1[$

$[0, 1[\times [0, 5] \times [0, 5] \times [4, 5]$ $[4, 5] \times [0, 1[\times [0, 5] \times [0, 5]$ $[4, 5] \times [4, 5] \times [0, 5] \times [0, 5]$

$[4, 5] \times [0, 5] \times [0, 5] \times [0, 1[$ $[4, 5] \times [0, 5] \times [0, 5] \times [4, 5]$ $[0, 5] \times [0, 1[\times [0, 1[\times [0, 5]$

$[0, 5] \times [0, 1[\times [4, 5] \times [0, 5]$ $[0, 5] \times [4, 5] \times [0, 1[\times [0, 5]$ $[0, 5] \times [4, 5] \times [4, 5] \times [0, 5]$

$[0, 5] \times [0, 5] \times [0, 1[\times [0, 1[$ $[0, 5] \times [0, 5] \times [0, 1[\times [4, 5]$ $[0, 5] \times [0, 5] \times [4, 5] \times [0, 1[$

$[0, 5] \times [0, 5] \times [4, 5] \times [4, 5]$

For instance, with $I = \{1, 2\}$, we have $[0, 1[\times [0, 1[\times [0, 5] \times [0, 5] \in R$, and therefore $[0, 5] \times [0, 5] \in \pi_{I^c}(\pi_I^{-1}([0, 1[\times [0, 1[))$. Because we have $[0, 1[\times [4, 5] \times [0, 5] \times [0, 5] \in R$, we have $[0, 1[\times [4, 5] \in \pi_I(R)$. However, $[0, 5] \times [0, 5] \notin \pi_{I^c}(\pi_I^{-1}([0, 1[\times [4, 5]))$ since $[0, 1[\times [4, 5] \times [0, 5] \times [0, 5] \notin R$. The processes $(p_1 \| p_2)$ and $(p_3 \| p_4)$ are thus not independent. With $I = \{1, 3\}$ the condition 3 of Algorithm 5.28 is satisfied and therefore the processes $(p_1 \| p_3)$ and $(p_2 \| p_4)$ are independent, and the program cannot be factorized further.

Finally, we should mention two extensions of these results. First, these can be generalized to programs of the form $p_1 \| p_2 \| \dots \| p_n$, where the threads p_i can contain any instruction except parallel composition (including conditional branchings and loops). Second, thanks to a result due to Ninin [134], the factorization can be efficiently performed, by exploiting the Boolean algebra structure of the collection of n-dimensional cubical regions. By abuse of language, we say that a partition $I_1, \dots, I_4$ of $[1 : n]$ is a factorization of X when the set of regions $\pi_{I_1}(X), \dots, \pi_{I_4}(X)$ is so.

Theorem 5.30 *Let $R \subseteq \mathbb{A}^n$ be the cubical cover, in normal form, of the complement of the state space X (in $\vec{I}^n$). The factorization of X is the finest partition of $[1 : n]$ whose elements are unions of subsets of the form $\left\{ i \mid \pi_i(u) \neq \vec{I} \right\}$, for $u \in R$.*

Example 5.31 Building on Example 5.29 the forbidden region of the program is

$$[1, 4[\times [0, 5] \times \underline{[1, 4[} \times [0, 5] \cup [0, 5] \times \underline{[1, 4[} \times [0, 5] \times \underline{[1, 4[}$$

The factorization immediately follows from Theorem 5.30, the associated partition of $[1 : 4]$ being $\{\{1, 3\}, \{2, 4\}\}$.

Chapter 6
The Category of Components

A major contribution of algebraic topology is to provide invariants of topological spaces up to homotopy, such as homotopy groups or homology groups. One of the simplest such invariants is the number of connected components of a space. Of course this invariant is very coarse since it does not distinguish between a disk and a circle, which both have one connected component:

A more refined invariant can be obtained by considering both the number of connected components and the number of homotopy classes of paths within each component i.e., the associated fundamental group. For instance, the disk has only one homotopy class of paths, while for the circle these homotopy classes are in bijection with $\mathbb{Z}$:

$$\ldots, -1, 0, 1, \ldots$$

From a more abstract point of view, these data can be obtained as the skeleton of the fundamental groupoid $\Pi_1(X)$ associated to a space X (see Definition 4.33). Here, taking the skeleton means considering the objects of the category up to isomorphism. The resulting category is thus the disjoint union of all the fundamental groups of the path-connected components of the space. Notice that this category, while retaining much information about the original space, is quite small compared to it. For instance, in the above examples, it has one object, which should be compared to the infinite number of points of the corresponding topological spaces.

Transposing this first nontrivial invariant to the setting of directed spaces, and defining a "directed component," is a real challenge. The nondirected case described

© Springer International Publishing Switzerland 2016
L. Fajstrup et al., *Directed Algebraic Topology and Concurrency*,
DOI 10.1007/978-3-319-15398-8_6

above suggests to us that we should begin by considering the fundamental category $\vec{\Pi}_1(X)$ associated to a d-space X, and "reduce" it in some way in order to obtain the *category of components* $\vec{\Pi}_0(X)$ of X. Two criteria that should be met by the definition of the category of components are that it should provide a category with few objects and morphisms, and that there is a functor $Q : \vec{\Pi}_1(X) \to \vec{\Pi}_0(X)$, from the fundamental category to the category of components, which induces a bijection on nonempty homsets (Theorem 6.23), i.e., it exhibits the same behavior as the original category.

Notice that the fundamental category of a d-space often has no nontrivial isomorphisms (this is typically the case for geometric models of programs), and therefore taking its skeleton will result in an isomorphic category, whereas we were aiming at making it smaller. Therefore, instead of identifying objects related by an isomorphism, we should use a stronger quotient, and identify objects related by an "inessential morphism": this notion should be a variant of the notion of isomorphism, which remains to be precisely defined. For instance, consider the directed space X depicted on the left, where directed paths are those going from left to right, obtained by gluing three copies of $\vec{I}$:

One easily gets convinced that a reasonable category of components associated to the topological space is the free category on the graph depicted on the right with three objects and two generators for morphisms: it does not really matter where we are, inside each interval $\vec{I}$. This is why we only need three objects in the category of components, and the arrows correspond to directed paths from one interval to the other.

Remark 6.1 This is more subtle than it appears: if one tries to construct component categories by identifying any two points related by a dipath, i.e., whenever there is a morphism in the category $\vec{\Pi}_1(X)$, the resulting category has only one object for our example. More generally, the category we obtain has one object for each zig-zag-connected component of the space and no nontrivial morphisms. On the contrary, if one identifies any two points x and y related both by a dipath from x to y and a dipath from y to x then the category we obtain is $\vec{\Pi}_1(X)$, which is not quotiented enough.

The category of components will provide a tool which will enable us to identify, in concurrent programs, actions which really have an impact on the execution of the program and forget about those which do not really matter up to commutation of actions, and thus provide a compact description of the fundamental category of a program, thus allowing for efficient exploration of its state space. For instance, consider the program p which is P_a;x:=1;$V_a \parallel P_a$;x:=2;V_a. Its geometric semantics $\check{G}_p$ is given on the left below:

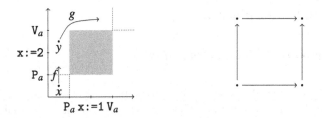

In some sense the execution corresponding to the directed path g was the "only thing" we could do starting from the point y: given another path g' starting from y, there exists paths h and h' such that $g \cdot h \sim g' \cdot h'$. In other words, up to dihomotopy, having performed g or not does not bring much information: in the end the variable x will contain the value 2. On the contrary, the path f corresponds to a real "choice" in the program (or more precisely its scheduler). It corresponds to the second process performing action $\mathtt{P}_a$: once this is done, the first process has no chance to perform the action $\mathtt{x} := \mathtt{1}$ first and, therefore, the variable x will have the value 2 in the end. If we are in a state where the second process has not performed action $\mathtt{P}_a$, it is still possible that x ends up with value 1. From a geometric point of view this corresponds to the fact that if we consider a path f' starting from x and going below the hole, there is no way to extend both f and f' in order to get two homotopic paths as before, because of the hole. The dotted lines in the figure delimit four connected regions of the space $\check{G}_p$. It can be checked that a directed path should be considered as inessential in the sense sketched above, if and only if it lies entirely in one of those regions. This explains why the associated category of components depicted on the right has four objects, and the arrows correspond to directed paths allowing to go from one region to the other, as earlier.

While the notion of the fundamental category is an immediate generalization in the directed setting of the notion of the fundamental groupoid, the definition of the category of components requires much more work and can be considered as a true novelty here: passing from a quotient w.r.t. every isomorphism to a quotient w.r.t. a suitable family of weak isomorphisms was not an easy step to perform. This notion was introduced in [49] and is now well understood for loop-free categories [86], but for more general categories, the right definition still eludes us. From now on, we therefore suppose given a (small) loop-free category $\mathscr{C}$, of which we will define the category of components:

Definition 6.2 A category $\mathscr{C}$ is *loop-free* if for every pair of objects $x, y \in \mathscr{C}$, $\mathscr{C}(x, y) \neq \emptyset$ and $\mathscr{C}(y, x) \neq \emptyset$ implies $x = y$ and $\mathscr{C}(x, x) = \{\mathrm{id}_x\}$.

The category $\mathscr{C}$ we consider is typically the fundamental category $\vec{\Pi}_1(\check{G}_p)$ of the geometric semantics of a program p without loops. Programs with loops (which give rise to non-loop-free categories) can be handled by either considering the universal dicovering of their geometric semantics, or their unrolling (see Remark 2.26). These can also be handled using the extensions mentioned in Sect. 6.4.

6.1   Weak Isomorphisms

6.1.1   Systems of Weak Isomorphisms

We begin by defining the class of weak isomorphisms (in the category $\mathscr{C}$). These contain and share many properties with isomorphisms, but are nontrivial for typical categories, such as the fundamental category of the geometric semantics of concurrent programs, even though those generally contain only trivial isomorphisms. As announced in the introduction, these will serve to identify objects when defining the category of components from the fundamental category. In order to do so, we should first observe some of the properties satisfied by isomorphisms.

Any morphism $f : x \rightarrow y$ in $\mathscr{C}$ induces, for every object $z \in \mathscr{C}$, a function $f^* : \mathscr{C}(y, z) \rightarrow \mathscr{C}(x, z)$ by precomposition: to a morphism $g \in \mathscr{C}(y, z)$ it associates $f^*(g) = g \circ f$. Similarly, it induces a function $f_* : \mathscr{C}(z, x) \rightarrow \mathscr{C}(z, y)$ for every object $z \in \mathscr{C}$ by post-composition.

Lemma 6.3 *A morphism $f : x \rightarrow y$ is an isomorphism if and only if the induced functions f^* and f_* are bijections for every object $z \in \mathscr{C}$.*

From a semantic perspective, an object y such that $\mathscr{C}(x, y) \neq \emptyset$ should be considered as part of the "future" of the object x, in the sense that there is a way of reaching y from x, and similarly x is in the "past" of y. It is thus reasonable that a morphism which does not change the future nor the past of any object is considered as inessential. We thus introduce a class of morphisms, which is wider than isomorphisms, by weakening the condition of the previous lemma.

Definition 6.4 A morphism $f : x \rightarrow y$ is a **weak isomorphism** if

- for every object $z \in \mathscr{C}$ such that $\mathscr{C}(y, z) \neq \emptyset$, the function $f^* : \mathscr{C}(y, z) \rightarrow \mathscr{C}(x, z)$ is a bijection, and
- for every object $z \in \mathscr{C}$ such that $\mathscr{C}(z, x) \neq \emptyset$, the function $f_* : \mathscr{C}(z, x) \rightarrow \mathscr{C}(z, y)$ is a bijection.

Example 6.5 Consider the category $\mathscr{C} = \vec{\Pi}_1(\check{G}_p)$ where p is the program already considered in the introduction of this chapter: $\mathrm{P}_a ; \mathtt{x:=1} ; \mathrm{V}_a \parallel \mathrm{P}_a ; \mathtt{x:=2} ; \mathrm{V}_a$. We are interested in the morphisms $f : x \rightarrow y$ and $f' : x' \rightarrow y'$ below.

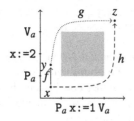 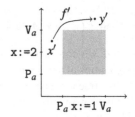

The morphism f on the left is not a weak isomorphism. Namely, we have $\mathscr{C}(y, z) \neq \emptyset$, for instance we have shown that the morphism $g : y \to z$ is in this set. However, the function $f^* : \mathscr{C}(y, z) \to \mathscr{C}(x, z)$ is not an isomorphism: in the directed space $\check{G}_p$, there is no way to extend the path f in order to obtain a path which will be homotopic to h. From a computing point of view this reflects the fact that the execution of P_a by the second process has an irreversible effect on the future: it will be no more possible for the first process to execute $x:=1$ first. The morphism f' shown on the right is a weak isomorphism. This corresponds intuitively to the fact that f' is, up to dihomotopy, the only possible path (i.e., execution of the program) to reach y' from x' (we should however underline that this intuition should not be taken too seriously in higher dimensions, for instance in the "floating cube," there is a unique total path which is not a weak isomorphism, see Sect. 6.2.3). Notice that the category $\mathscr{C}$ has nontrivial weak isomorphisms (such as f'), whereas the only isomorphisms are identities.

In this example, we have seen that a weak isomorphism corresponded, from a computing point of view, to the only way of executing a program up to homotopy. Namely,

Lemma 6.6 *Given a loop-free category $\mathscr{C}$, if $f : x \to y$ is a weak isomorphism then f is the only morphism from x to y.*

Proof Let $f : x \to y$ be a weak isomorphism. The homset $\mathscr{C}(x, x)$ is a singleton because $\mathscr{C}$ is loop-free and f_* is a bijection from $\mathscr{C}(x, x)$ to $\mathscr{C}(x, y)$. $\qquad\square$

We have illustrated why weak isomorphisms are a very reasonable notion. However, there are still too many of these. In order to convince the reader, consider the fundamental category of the geometric semantics of the Swiss flag program, see Example 4.21:

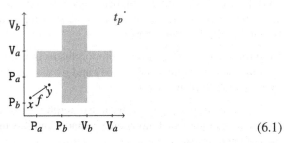

$$\tag{6.1}$$

Notice that the terminal position t_p can be reached from point x, but not from point y: in other words, the point y is doomed whereas x is not. Consequently, the morphism f (drawn in the picture above) should not be considered as inessential. This illustrates why, in order to rule out such morphisms, we need to impose further restrictions upon the class of weak isomorphisms, making it closer to the class of isomorphisms. With this goal in mind, we notice that isomorphisms are stable under pushouts and pullbacks:

Lemma 6.7 *Given an isomorphism $f : x \to y$, for every coinitial morphism $g : x \to z$ there exists a pushout*

$$
\begin{array}{ccc}
 & g' \nearrow \ \nwarrow f' & \\
y & & z \\
 & f \nwarrow \ \nearrow g & \\
 & x &
\end{array}
$$

and the morphism f' obtained as the pushout of f along g is also an isomorphism, and similarly for pullbacks w.r.t. cofinal morphisms.

In the following, we will thus be interested in weak isomorphisms which satisfy similar stability properties and we will consider as inessential the morphisms in the maximal system, w.r.t. inclusion, of weak isomorphisms:

Definition 6.8 A *system of weak isomorphisms* is a collection Σ of weak isomorphisms of $\mathscr{C}$, which is stable under pushouts and pullbacks, and contains all isomorphisms.

Example 6.9 In the geometric semantics of the Swiss flag, a collection of the morphisms containing the morphism f of (6.1) cannot be stable under pushouts since there is no pushout with any path $g : x \twoheadrightarrow t_p$.

6.1.2 A Maximal System

Since the composite of weak isomorphisms is still a weak isomorphism, it is easy to see that given a system of weak isomorphisms, its closure under composition is still a system of weak isomorphisms. We will therefore suppose that the systems we consider are closed under composition, and we denote by $\mathrm{SWI}(\mathscr{C})$ the collection of all such systems. When the category $\mathscr{C}$ is loop-free, those systems can be shown to be pure, in the following sense.

Definition 6.10 A collection Σ of morphisms is said to be *pure* when for every pair of composable morphisms $f : x \to y$ and $g : y \to z$, if their composite $g \circ f$ belongs to Σ then so do both f and g.

An element of a system of weak isomorphisms typically corresponds to an execution trace along which no choice is made, and it is thus natural to expect that any part of such an execution trace also satisfies this property. The following lemma shows that it is indeed the case. This result moreover plays a key role in the proof of Proposition 6.12.

Lemma 6.11 *Given a loop-free category $\mathscr{C}$ and $\Sigma \in \mathrm{SWI}(\mathscr{C})$, the collection Σ is pure.*

The set $\mathrm{SWI}(\mathscr{C})$ is partially ordered by inclusion thus giving rise to a complete lattice: greatest lower bound of two systems is given by the set-theoretic intersection, while least upper bound is the least system containing the set-theoretic union.

Proposition 6.12 *The poset* SWI($\mathscr{C}$) *ordered by inclusion forms a complete lattice.*

A maximal system of weak isomorphisms thus exists and will be the main object of our attention.

This system can be obtained as an application of the Knaster–Tarski fixpoint theorem [158] as follows. If we write W for the set of all weak isomorphisms of $\mathscr{C}$, the set $\mathfrak{P}(W)$ of subsets of W ordered by inclusion is a lattice with W as maximal element. Given $\Sigma \in \mathfrak{P}(W)$, we write $\Phi(\Sigma)$ for the collection of morphisms $f \in \Sigma$ such that for every coinitial (resp. cofinal) morphism $g \in \mathscr{C}$, the pushout (resp. pullback) of f along g exists and belongs to Σ. The function $\Phi : \mathfrak{P}(W) \to \mathfrak{P}(W)$ thus defined is clearly order-preserving and its greatest fixpoint can be obtained as the limit of the sequence $\Phi^n(W)$. This characterization of the greatest system of weak isomorphisms can be used in order to compute it algorithmically for loop-free categories which are finitely presented, such as the fundamental category of the cubical semantics of a loop-free concurrent program. In the following, we will use it to reduce the size of the category $\mathscr{C}$ by turning the morphisms in this system into identities.

6.1.3 Quotienting by Weak Isomorphisms

Suppose given a class Σ of morphisms of a category $\mathscr{C}$, and we would like to somehow "remove" the morphisms in Σ from $\mathscr{C}$. There are two categorical constructions available in order to do so. We can either force them to be identities by constructing the quotient category $\mathscr{C}/\Sigma$, or force them to be isomorphisms by constructing the localization $\mathscr{C}\Sigma^{-1}$ of $\mathscr{C}$ w.r.t. Σ. We study the first construction in this section and will consider localization in the next one.

Definition 6.13 The *quotient* of a category $\mathscr{C}$ by Σ consists of a category $\mathscr{C}/\Sigma$, together with a functor $Q : \mathscr{C} \to \mathscr{C}/\Sigma$ sending morphisms in Σ to identities and called the *quotient functor*, such that a functor $F : \mathscr{C} \to \mathscr{D}$ factors uniquely through Q if and only if it sends all morphisms in Σ to identities.

$$\begin{array}{ccc} \mathscr{C} & \xrightarrow{\ F\ } & \mathscr{D} \\ {\scriptstyle Q}\downarrow & \nearrow & \\ \mathscr{C}/\Sigma & & \end{array}$$

Such a category always exists, see [8].

Example 6.14 Given a d-space X, consider its fundamental category $\mathscr{C}$ and Σ the collection of all morphisms of $\mathscr{C}$. The category $\mathscr{C}/\Sigma$ is the discrete category whose objects are the path-connected components of the space X.

Example 6.15 Consider the category associated to the monoid $(\mathbb{N}, +, 0)$: it has one object, its morphisms are integers, and composition is given by addition. We will continue to use $\mathbb{N}$ to denote this category. Given the set $\Sigma = \{1\}$ of morphisms, the quotient category $\mathbb{N}/\Sigma$ is the terminal category with one object and only the identity on this object as morphism.

Example 6.16 Consider the category $\mathscr{C}$ with two objects x and y and two nontrivial arrows $f, g : x \to y$, and $\Sigma = \{g\}$. The quotient category $\mathscr{C}/\Sigma$ is isomorphic to $\mathbb{N}$, as defined in Example 6.15.

From now on, we suppose that Σ is a fixed system of weak isomorphisms in a loop-free category $\mathscr{C}$. We look in detail at the corresponding quotient categories and in particular, the case where Σ is the maximal such system will give us our definition of components:

Definition 6.17 The **category of components** $\vec{\Pi}_0(\mathscr{C})$ associated to a loop-free category $\mathscr{C}$ is the quotient category $\mathscr{C}/\Sigma$ where Σ is the greatest system of weak isomorphisms of $\mathscr{C}$. Given a directed topological space X, we simply write $\vec{\Pi}_0(X)$ for $\vec{\Pi}_0(\vec{\Pi}_1(X))$.

Remark 6.18 We will also consider the case where Σ is not maximal in Sect. 6.3. In this case, the category $\mathscr{C}/\Sigma$ is called a *category of precomponents* and can be thought of as an approximation of the category of components, in the sense that the latter is a quotient of the former.

We say that two objects x and y are Σ *-connected* when they can be joined by a zig-zag of morphisms in Σ, i.e., there exists a finite sequence $x_0, \ldots, x_n$ of objects of $\mathscr{C}$ such that $x_0 = x, x_n = y$ and for every index i, there is a morphism $f : x_i \to x_{i+1}$ or a morphism $f : x_{i+1} \to x_i$ in Σ. This defines an equivalence relation on objects of $\mathscr{C}$ whose equivalence classes are called the Σ-*components* of $\mathscr{C}$.

These components can be shown to be structured as follows. Given a Σ-component K of $\mathscr{C}$, consider the full subcategory $\mathscr{K}$ of $\mathscr{C}$ whose objects are the elements of K. By Lemma 6.6, there is at most one morphism between two objects of $\mathscr{K}$, i.e., the category is a preorder, and since $\mathscr{C}$ is supposed to be loop-free this preorder is actually a partial order.

Proposition 6.19 *Let Σ be a system of weak isomorphisms of a loop-free category $\mathscr{C}$. Suppose given a Σ-component K of $\mathscr{C}$. The following can be shown:*

- *the relation $\preceq$ on K defined by $x \preceq y$ whenever there is a morphism $f : x \to y$ in $\mathscr{C}$ is a partial order,*
- *the resulting poset $(K, \preceq)$ is isomorphic to the full subcategory of $\mathscr{C}$ whose objects are the elements of K,*
- *the poset $(K, \preceq)$ is a lattice.*

If moreover Σ is the maximal system, a category $\mathscr{C}$ is a component (i.e., its objects are all in the same Σ-component) if and only if $\mathscr{C}$ is a lattice.

Two objects x and y which are Σ-connected thus admit a greatest lower bound $x \wedge y$ (i.e., a coproduct) and a least upper bound $x \vee y$ (i.e., a product) in the category $\mathscr{K}$ corresponding to their component as above. Moreover, there is a unique morphism from x or y to $x \vee y$, and from $x \wedge y$ to x or y, and this morphism belongs to Σ. In the category $\mathscr{K}$, it is easy to see that the square

$$(6.2)$$

is both a pushout and a pullback. It turns out that this property is preserved by the embedding of $\mathcal{K}$ into $\mathcal{C}$:

Proposition 6.20 *Given a Σ-component $\mathcal{K}$, the inclusion functor $\mathcal{K} \hookrightarrow \mathcal{C}$ preserves pushouts and pullbacks, i.e., the image of a square (6.2) is both a pushout and a pullback.*

The above proposition is often quite useful to compute pushouts and pullbacks. Namely, it implies that

Corollary 6.21 *Given two arrows $f : x \to y$ and $g : x \to z$ in Σ, their pushout exists and is given by the arrows $y \to y \vee z$ and $z \to y \vee z$, and dually for pullbacks.*

We now describe the quotient category $\mathcal{C}/\Sigma$ in the particular case when Σ is a system a weak isomorphisms of a loop-free category $\mathcal{C}$. This covers in particular the construction of the category of components. Given an object x of $\mathcal{C}$, we write $[x]$ for its Σ-component. Suppose given two morphisms $f : x \to y$ and $f : x' \to y'$ of $\mathcal{C}$ such that x and x' are Σ-connected, and y and y' are Σ-connected. We say that f and f' are Σ-*equivalent* when the diagram

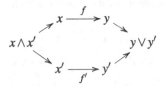

commutes. This defines an equivalence relation on the morphisms of $\mathcal{C}$ and we write $[f]$ for the equivalence class of a morphism f of $\mathcal{C}$. These equivalence classes enable us to provide an easy description of the quotient category:

Theorem 6.22 *The quotient category $\mathcal{C}/\Sigma$ is (isomorphic to) the category whose objects are of the form $[x]$ for some object x of $\mathcal{C}$, morphisms are of the form $[f]$ for some morphism f of $\mathcal{C}$, composition is given by composition in $\mathcal{C}$, and identities are equivalence classes of those in $\mathcal{C}$.*

The above category can be shown to be well defined. Namely, composition in $\mathcal{C}$ is easily checked to be compatible with Σ-equivalence. Moreover, suppose given two morphisms $f : x \to y$ and $g' : y' \to z'$ in $\mathcal{C}$ such that y and y' are Σ-connected. These induce morphisms $[f] : [x] \to [y]$ and $[g'] : [y'] \to [z']$ in the quotient category, with $[y] = [y']$. We can find composable representatives in the classes $[f]$ and $[g]$ using Proposition 6.19 and the stability of Σ under pushouts:

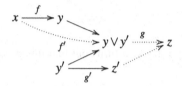

Above, the morphism f' is obtained by composing f with the morphism $y \to y \vee y'$, and g as the pushout of g' along the morphism $y' \to y \vee y'$. It can easily be shown that f and f' are Σ-equivalent, as well as g and g'.

The quotient functor $\mathscr{C} \to \mathscr{C}/\Sigma$ is of course the functor sending an object x of $\mathscr{C}$ to $[x]$ and a morphism f to $[f]$. From this, one can draw many interesting simple observations. For instance, the quotient functor is surjective on morphisms, and if $\mathscr{C}$ is finite then so is $\mathscr{C}/\Sigma$. These properties are not true in the general case (for an arbitrary set Σ), for instance, Example 6.16 provides a counterexample to both of them. One can also show the following fundamental property, which ensures that the quotient category behaves locally as the original category:

Theorem 6.23 *Given two objects x and y of $\mathscr{C}$ such that $\mathscr{C}(x, y) \neq \emptyset$, the function $\mathscr{C}(x, y) \to \mathscr{C}/\Sigma([x], [y])$ induced by the quotient functor is a bijection.*

Proof Suppose given a morphism of $\mathscr{C}/\Sigma([x], [y])$. This morphism is of the form $[f']$ for some morphism $f' : x' \to y'$ of $\mathscr{C}$ with $x' \in [x]$ and $y' \in [y]$. By Proposition 6.19, we therefore have a diagram of the following form, without the dotted arrows:

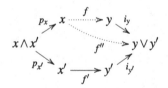

Because we have supposed $\mathscr{C}(x, y) \neq \emptyset$, we have $\mathscr{C}/\Sigma([x], [y]) \neq \emptyset$ and therefore $\mathscr{C}/\Sigma([x], [y \vee y']) \neq \emptyset$, by post-composing with i_y. Since the morphism p_x is in Σ, it is a weak isomorphism and therefore, since $\mathscr{C}/\Sigma([x], [y \vee y']) \neq \emptyset$, there exists a unique morphism $f'' : x \to y \vee y'$ such that $f'' \circ p_x = i_{y'} \circ f' \circ p_{x'}$. Similarly, because $i_y \in \Sigma$ and $\mathscr{C}/\Sigma([x], [y]) \neq \emptyset$, we have a unique morphism $f : x \to y$ such that $i_y \circ f = f''$. We therefore have $[f] = [f']$ and f is the unique possible pre-image of $[f']$ under the quotient functor.   □

Remark 6.24 If we had taken Σ to be a system of morphisms (i.e., a set of morphisms containing isomorphisms and stable under composition, pushouts, and pullbacks), without supposing the morphisms to be weak isomorphisms, Theorem 6.23 would not be satisfied in general. For instance, consider the category $\mathscr{C}$ which is the fundamental category of the asynchronous graph on the left below:

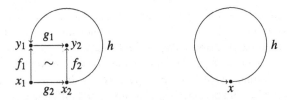

It is easy to see that the class of morphisms $\Sigma = \{f_1, g_1, f_2, g_2, g_1 \circ f_1, g_2 \circ f_2\}$ is a system of morphisms and therefore the quotient category $\mathscr{C}/\Sigma$ is the free category on the graph on the right above, i.e., it is the category $\mathbb{N}$ (this is a small variation of Example 6.16). Since the category $\mathscr{C}$ is finite and the set $\mathscr{C}/\Sigma(x, x)$ is not Theorem 6.23 clearly does not hold. This shows the importance of considering weak isomorphisms when defining our quotient.

6.1.4 Other Definitions

The quotient construction of $\mathscr{C}/\Sigma$, described in the previous section, amounts to turning the morphisms of Σ into identities. Another possibility consists in converting them into isomorphisms, thus giving rise to the notion of the localization $\mathscr{C}[\Sigma^{-1}]$ of $\mathscr{C}$ w.r.t. Σ. This notion is much more common than quotients, because in category theory, constructions are usually performed up to isomorphism. We recall here its definition and explain how it relates to the earlier quotient construction.

Definition 6.25 The *localization* of a category $\mathscr{C}$ with respect to Σ consists of a category $\mathscr{C}[\Sigma^{-1}]$, together with a functor $L : \mathscr{C} \to \mathscr{C}[\Sigma^{-1}]$ sending morphisms in Σ to isomorphisms, such that a functor $F : \mathscr{C} \to \mathscr{D}$ factors uniquely through L if and only if it sends all morphisms in Σ to isomorphisms.

$$\begin{array}{ccc} \mathscr{C} & \xrightarrow{F} & \mathscr{D} \\ {\scriptstyle L}\downarrow & \nearrow & \\ \mathscr{C}[\Sigma^{-1}] & & \end{array}$$

We refer to [13] for more details about this notion, which can be shown to exist for any $\mathscr{C}$ and Σ. When the set Σ of morphisms satisfies certain well-known properties (see *calculus of fractions* in [13]), which are satisfied by our systems of weak isomorphisms, the localization admits a nice description as a *category of fractions* with a neater definition of the morphisms as "fractions" gf^{-1} with $f \in \Sigma$.

Example 6.26 Consider again the category $\mathbb{N}$ corresponding to the monoid of integers introduced in Example 6.15, with $\Sigma = \{1\}$. The localization $\mathbb{N}[\Sigma^{-1}]$ is the category with one object, the set of morphisms is $\mathbb{Z}$, and composition is given by addition. Notice that $\mathbb{N}[\Sigma^{-1}]$ is *not* isomorphic or even equivalent to the quotient category $\mathbb{N}/\Sigma$ which is the terminal category, as described in Example 6.15.

The previous example illustrates the fact that we generally do not have an equivalence of categories between $\mathscr{C}/\Sigma$ and $\mathscr{C}[\Sigma^{-1}]$ if we do not suppose that $\mathscr{C}$ is loop-free. The following question thus arises: what would have happened if we had defined the category of components as $\mathscr{C}[\Sigma^{-1}]$ with Σ some system of weak isomorphisms of the category $\mathscr{C}$? Fortunately, in the case where $\mathscr{C}$ is loop-free, we have an equivalence of categories, i.e., it would not have made a substantial difference.

Suppose that Σ is a system of weak isomorphisms for the loop-free category $\mathscr{C}$ and consider the functor $Q : \mathscr{C} \to \mathscr{C}/\Sigma$. By definition, Q sends morphisms in Σ to identities and thus to isomorphisms. By definition of the localization, there thus exists a functor $P : \mathscr{C}[\Sigma^{-1}] \to \mathscr{C}/\Sigma$ such that $P \circ L = Q$.

Theorem 6.27 *The functor $P : \mathscr{C}[\Sigma^{-1}] \to \mathscr{C}/\Sigma$ is an equivalence of categories.*

It was explained in the introduction that the construction of the category of components can be thought of as a directed variant of taking the skeleton of a groupoid. Following this analogy, we should be able to exhibit $\vec{\Pi}_0(\mathscr{C})$ as a full subcategory of $\mathscr{C}$ which meets each component exactly once. In other words, we would like to provide the quotient functor $Q : \mathscr{C} \to \mathscr{C}/\Sigma$ with a section, i.e., a functor $S : \mathscr{C}/\Sigma \to \mathscr{C}$ such that $Q \circ S$ is the identity functor on $\mathscr{C}/\Sigma$. This motivates the following definition.

Definition 6.28 A *choice function j* is a function from the Σ-components of $\mathscr{C}$ to the objects $\mathscr{C}$ such that for every Σ-component K the object $j(K)$ belongs to K. Such a choice function is *admissible* when for every pair of Σ-components K and K', if there exists $x \in K$ and $x' \in K'$ such that $\mathscr{C}(x, x') \neq \emptyset$ then $\mathscr{C}(j(K), j(K')) \neq \emptyset$.

If the number of components is finite, an admissible choice function always exists.

Proposition 6.29 *If the number of Σ-components is finite, then there exists an admissible choice function.*

Proof We write $\mathscr{C}_0$ for the set of objects of $\mathscr{C}$ and $\mathscr{K}$ for the set of Σ-components of $\mathscr{C}$. We suppose given a function $i : \mathscr{K} \to \mathfrak{P}(\mathscr{C}_0)$, which to every component K associates a set of objects such that $i(K) \subseteq K$ and for every pair of components $K, K' \in \mathscr{K}$, if there exists $(x, x') \in K \times K'$ such that $\mathscr{C}(x, x') \neq \emptyset$, then there exists $(y, y') \in i(K) \times i(K')$ such that $\mathscr{C}(y, y') \neq \emptyset$. Such a function clearly exists and we can suppose that $i(K)$ is finite for every component K, because the number of (pairs of) components is supposed to be finite. The function j defined by $j(K) = \bigvee i(K)$ can then be shown to be an admissible choice function. Notice that the least upper bound of $i(K)$ exists because the component K is a lattice by Proposition 6.19, and there are finitely many components. $\qquad\square$

Theorem 6.30 *Given an admissible choice function j, the quotient category $\mathscr{C}/\Sigma$ is isomorphic to the full subcategory of $\mathscr{C}$ whose objects are those in the image of j.*

We could thus, alternatively, have defined the category of components as the full subcategory on the image of an admissible choice function. Interestingly, this last definition is the only one to give satisfactory results with loops, as mentioned in Sect. 6.4.

Example 6.31 Consider the fundamental category associated to the geometric semantics of the program $P_a ; V_a \parallel P_a ; V_a$, which has four components K_1, K_2, K_3, and K_4 as shown on the left below:

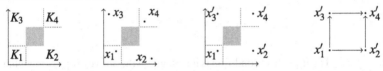

We can build a choice function i by picking an arbitrary point x_k in each component K_k, as shown on the middle left, and defining $i(K_k) = x_k$. However, the resulting function will not be admissible in general: in our example, there is no morphism (i.e., directed path) from x_1 to x_2, because x_2 is "below" x_1 (hence x_2 is not comparable to x_1 using the componentwise partial ordering), whereas some points of K_2 can be reached from some points in K_1. The proof of Proposition 6.29 provides a way to construct an admissible choice function j from the choice function i; on the middle right, we have illustrated points x_i' such that $j(K_k) = x_k'$. The full subcategory with the points x_k' as objects is drawn on the right, and is the category of components by Theorem 6.30.

6.2 Examples of Categories of Components

6.2.1 Trees

We now investigate the categories of components associated to the fundamental categories of trees (or of their geometric realization).

Definition 6.32 A *tree* T is a graph which admits a vertex x_0 called its *root* such that for every vertex x there is a unique path $x_0 \twoheadrightarrow x$. A vertex of a tree is called a *leaf* when it is the source of no edge and a *branching* when it is the source of at least two distinct edges.

The fundamental category of the category generated by such a tree can be characterized as follows.

Proposition 6.33 *Suppose given a finite tree T and write T^* for the category generated by T. The categories of components $\vec{\Pi}_0(T^*)$ and $\vec{\Pi}_0(|T|)$ are both isomorphic to the full subcategory of T^*, or of $\vec{\Pi}_1(|T|)$, whose objects are the branching vertices and the leaves.*

Example 6.34 Consider the tree depicted on the left. We have circled the vertices which are either branchings or leaves. The category of components of T^* or of $\vec{\Pi}_1(|T|)$ is isomorphic to the free category on the graph depicted on the right.

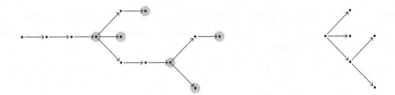

Remark 6.35 The finiteness condition is necessary in the above proposition. Namely, consider the (infinite) tree with $\mathbb{N}$ as set of vertices and for every $n \in \mathbb{N}$ there is an edge from n to $n + 1$. There are no branching vertices or leaves. However, the associated fundamental category is reduced to an object, i.e., it is the terminal category.

The simple characterization given for trees does not directly generalize to directed acyclic graphs. For instance, the graph on the left admits the free category on the graph on the right as category of components.

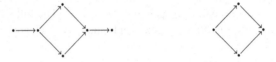

A path in a graph is *non-branching* when its vertices are the source of at most one edge and the target of at most one edge, except maybe the source and the target of the path. It can be shown that the directed components of a directed acyclic graph are in bijection with the maximal non-branching paths of G.

6.2.2 Cubical Regions in Dimension 2

Consider a cubical region X in dimension 2. Since the set of cubical regions is closed under complement (see Sect. 5.1), this space is of the form $X = \vec{I}^2 \setminus \bigcup_{i=1}^{l} R^i$ with $R^i = [x_1^i, y_1^i[\times [x_2^i, y_2^i[$ (where "[" is either "[" or "]"), and is typically generated as the geometric semantics of a simple program in dimension 2, see Sect. 4.1.4. The category of components $\vec{\Pi}_0(X)$ of the associated fundamental category can be computed as follows. For every lower left corner $x^i = (x_1^i, x_2^i)$ of a forbidden rectangle, draw a horizontal straight line starting from x and extended to the left until a dead point is met (or the boundary of $\vec{I}^2$), as well as a vertical line extending x downward until a dead point. Similarly, extend every upper right corner $y = (y_1^i, y_2^i)$ upward vertically and forward horizontally. The connected components of the space obtained from X by removing those lines are the components of X. When a line splits two neighboring components, whether its points belong to a component or the other depend on whether the boundaries of the rectangle R^i from which it originates are open or closed. There is a morphism from one component to another when there is a directed path from a point of the first to a point of the second, and two morphisms are equal when the corresponding paths are dihomotopic.

Example 6.36 We have shown below four such spaces, as well as (superimposed) the asynchronous graph (see Definition 3.3.2) whose fundamental category is the category of components of the space.

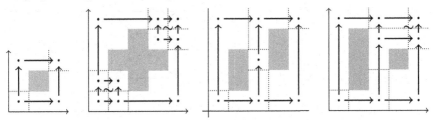

An algorithm based on similar ideas, generalized to any dimension, will be presented in Sect. 6.3.

6.2.3 The Floating Cube and Cross

Consider the program $p = \mathtt{P}_a \, ; \mathtt{V}_a \, \| \, \mathtt{P}_a \, ; \mathtt{V}_a \, \| \, \mathtt{P}_a \, ; \mathtt{V}_a$ with $\kappa_a = 2$. It has geometric semantics $\check{G}_p = \vec{I}^3 \backslash \,]\frac{1}{3}, \frac{2}{3}[^3$ which is a "floating cube," as already mentioned in Example 4.15. The category of components associated to its fundamental category $\vec{\Pi}_0(\check{G}_p)$ has 26 objects and is in fact isomorphic to the fundamental category associated to its cubical semantics $\vec{\Pi}_1(\check{C}_p)$, which was shown in (3.5) at the beginning of Sect. 3.4 (see also the figure on the right below).

This can be shown as follows. Considering the hyperplane depicted on the left below, we see that the morphism f_1 is not a weak isomorphism, for the same reasons as the morphism f in Example 6.5. Moreover, the morphism f_0 depicted in the second picture below, from the left, has no pushout along the morphism g, because of the presence of the hole (we have pictured two incomparable ways to close the span (f_0, g) as a commutative square). Furthermore, on the third picture, one readily checks that f_1 is the pullback of f_i, with $i \in \{2, \dots, 5\}$, along with an appropriate morphism. Hence, none of the morphisms $f_1, \dots, f_5$ belongs to a system of weak isomorphisms because the latter is stable under pullbacks and has been proven not to contain f_1.

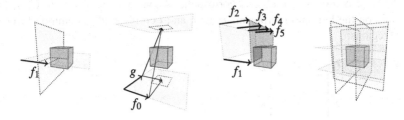

If we consider the same program p, but in the case where a is a mutex, the geometric semantics is shown on the left, and the corresponding category of components has 14 objects, and is freely generated by the graph shown on the right:

6.3 Computing Components

In the case of the geometric semantics of simple programs, it can be observed that the categories of components can always be presented as the fundamental category of an asynchronous graph (i.e., a two-dimensional precubical set, see Definition 3.32), as illustrated in the previous section (see Example 6.36 for instance). It is quite difficult to compute such an asynchronous graph in general. However, we can easily compute an asynchronous graph whose fundamental category corresponds to quotienting the geometric semantics by some nontrivial system of weak isomorphisms, which is not the maximal one in general. We briefly describe in this section an algorithm comput-ing such an approximation of the category of components of a simple program: given a program p, the algorithm will compute a category which is a quotient of $\vec{\Pi}_1(\check{G}_p)$, of which the category of components $\vec{\Pi}_0(\check{G}_p)$ is a quotient, and both quotients trivialize weak isomorphisms only. In other words, it computes a category of precomponents, in the sense of Remark 6.18.

6.3.1 The Case of One Hole

Recall from Sect. 6.2.2 that the geometric semantics of the simple program P_a; $V_a \parallel P_a$; V_a, where a is a mutex, is obtained by removing a square (named R below) from a square, and the associated component category is the skeletal category pictured on the right below, whose objects correspond to the four regions A, B, C, and D pictured on the left, and morphisms f, g, h, and i correspond to common faces of these regions:

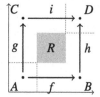
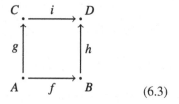

$$(6.3)$$

Two interesting remarks can be made on this example. First, the category of components is the free category on the graph above with four vertices and four edges, and is thus the fundamental category of an asynchronous graph (with trivial dihomotopy relation). Second, the regions can be obtained by separating the space by the dotted lines which originate from the lowest and highest point of the forbidden square.

As another example, consider the category of components of a program consisting of n copies of the program $P_a ; V_a$ in parallel, where the resource a is of capacity $n-1$. It can be computed using a generalization of the computation for the floating cube of Sect. 4.3.1, which is the particular case where $n = 3$. Similar remarks can be formulated. First, for $n = 3$, the category of components is the fundamental category of the two-dimensional precubical set with 26 vertices, 32 edges, and 24 squares (or 2-cells), which describes the (cubical) barycentric subdivision of a simple hollow cube (8 vertices, 8 edges, 6 squares). Second, the regions are obtained by separating the space using 6 planes, each of which is following one of the 6 square faces of the forbidden cube.

From these examples, one can build the following intuition for the general n-dimensional case. There are two cases to consider.

- Consider the space $X_{n,n}$ which is the complement of a hypercube of dimension n (the forbidden region), in the interior of a bigger hypercube of dimension n. This space is produced when n processes ask for locks on a resource of capacity $n - 1$. Then the regions that will become objects in the category of components are the $3^n - 1$ regions delimited by the 2^n hyperplanes containing the 2^n faces of the forbidden region, minus the forbidden region. The unique morphism from a region to the neighboring one is in bijection with their common face. Four regions that intersect create relations between the morphisms from one of these four components to some of its neighboring components.

- In analogy with the Swiss flag (see Example 4.21), consider the space $X_{n,n-k}$ which is the complement of a cylinder with hypercube section of dimension $n - k$, going all though some bigger hypercube of dimension n. This type of hole arises when considering at least $n - k$ among n processes trying to lock a resource of capacity $n - k - 1 < n - 1$ (many of these holes can be created by the semantics). All these hypercubes are as usual isothetic hypercubes, so $X_{n,n-k}$ is isomorphic to $I^k \times X_{n-k,n-k}$. As for the fundamental category, components commute with the Cartesian product, and hence the maximal component category is the one of $X_{n-k,n-k}$ (with $3^{n-k} - 1$ regions).

6.3.2 The General Case

Suppose we want to determine now a category of components for a simple program where the forbidden area is generated by multiple hypercubes, such as the program $p = P_a ; V_a ; P_b ; V_b \| P_b ; V_b ; P_a ; V_a$ where a and b are mutexes. Its directed geometric semantics is obtained from the one of $P_a ; V_a \| P_a ; V_a$ studied in the previous section by carving a new square hole, as shown on the left of (6.4), and the corresponding category of components is the fundamental category of the asynchronous graph shown on the right, which corresponds to the subdivision of the space shown in the middle.

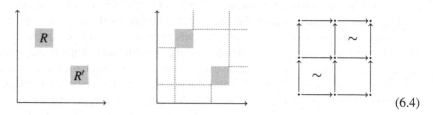

$$(6.4)$$

If we start from only the hole R, as covered in the previous section, carving a new hole such as R' will create new hyperplanes from the lowest and highest points of the hole (the dotted lines in the middle of (6.4)), which will cut the components we had at the previous stage into new components. Moreover, these will cut new segments of hyperplanes, and hence will generate new morphisms from one component to neighboring ones, as we sketched in Sect. 6.3.1. When adding R', an interesting situation can be observed: the fact that two of these hyperplanes can intersect in a point produces two pairs of composable morphisms with the same start and end regions. A simple argument shows that the corresponding composites should commute; hence an independence tile should appear in the corresponding presentation using asynchronous graphs. The intuition now is that regions of codimension 0 will form points in the asynchronous graph corresponding to the category of components we are describing, regions of codimension 1 (segments of hyperplanes) will form edges, and regions of codimension 2 (points in general) will form independence tiles.

If we write X_R for the space in the case where there is only the hole R, we know a presentation of the associated category of components $\vec{\Pi}_0(X_R)$: following (6.3), it is presented by the two-dimensional precubical set C_R which contains four objects A, B, C, and D; four morphisms f, g, h, and i; and no 2-cell. The case of the space $X_{R'}$ containing only the hole R' is obviously similar, and we use similar primed notations. The elements of the category of components shown on the right of (6.4) can be seen as pairs of elements of each category of components, a pair consisting of elements of dimension k and l is of dimension $k + l$, as illustrated below. For instance (f, g') consists of two 1-cells and is thus a 2-cell (a tile), and similarly (A, A') consists of two 0-cells and is thus a 0-cell (a vertex).

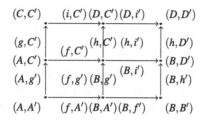

It can be noticed that those cells are elements of the tensor product $C_R \otimes C_{R'}$ of the two precubical sets. However, not every cell in the tensor product is part of the above presentation: a simple criterion characterizing the cells which are present can be formulated as follows. Each cell in C_R corresponds to a portion of the space X_R as explained above: a 0-cell A corresponds to a region $|A|$, a 1-cell f to a segment $|f|$, and a 2-cell I to a point $|I|$, and similarly for $C_{R'}$. A 0-, 1- or 2-cell $(x, x') \in C_R \otimes C_{R'}$ is part of the above cubical set precisely when $|x| \cap |x'| \neq \emptyset$. For instance, the cell (f, g') is present because the segments $|f|$ and $|g'|$ intersect at a point, whereas (g, f') is not present because the corresponding segments do not intersect.

The above observation leads to a general procedure for computing a presentation of a category of precomponents for the geometric semantics of a simple program:

Algorithm 6.37 Suppose given a space of the form $X = \vec{I}^n \setminus \bigcup_{i=1}^{l} R^i$ where the R^i are cubical regions, a typical geometric semantics of a simple program of dimension n.

1. If $l = 1$, i.e., there is only one forbidden region R^1, a presentation of a category of components can be computed as explained in Sect. 6.3.1.
2. Otherwise,

   a. choose l' such that $1 \leq l' < l$ (typically $l' = l/2$),
   b. recursively compute two-dimensional precubical sets C and C', respectively, presenting categories of precomponents of $\vec{I}^n \setminus \bigcup_{i=1}^{l'} R^i$ and $\vec{I}^n \setminus \bigcup_{i=l'+1}^{l} R^i$,
   c. return the precubical subset of $C \otimes C'$ consisting of 0-, 1-, and 2-cells (x, x') such that $|x| \cap |x'| \neq \emptyset$.

We insist on the fact that it gives a quotient category which is in general less quotiented than the category of components (as illustrated in Example 6.40): we get a nontrivially "compressed" state space which might not be as optimal as the category of components, but which is useful for static analysis as we exemplify briefly below. We refer the reader to [67] for more details, and to [71] for formal relationships between components and state-space reduction techniques, as used in model checking (e.g. , persistent sets).

Example 6.38 (Swiss flag) Consider the Swiss flag Example 3.22. Its geometric semantics is shown on the left below (see Example 4.21), together with its 10 components. The category of components is presented by the asynchronous graph on the right. Notice that we have two asynchronous tiles corresponding to the commutation relations $f'_2 \circ f'_1 = f_2 \circ f_1$ and $g'_2 \circ g'_1 = g_2 \circ g_1$.

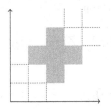
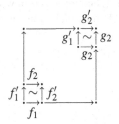

Example 6.39 (*Dining philosophers*) In the case of the 2 and 3 dining philosophers Example 3.23, the category of components is presented by the precubical sets below, where the filled surfaces represent 2-cells (i.e., commutation tiles).

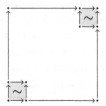

By Theorem 6.23, we know that we can deduce from any category of precomponents the maximal morphisms, i.e., the equivalence classes of maximal execution traces. In the case of the 2 dining philosophers, we find 3 maximal traces, 2 of which are non-deadlocking, and in the case of the 3 dining philosophers, we find 7 maximal traces, 6 of which are non-deadlocking. More generally, the n dining philosophers exhibit $2^n - 1$ execution traces up to dihomotopy, one of which is deadlocking.

Example 6.40 This algorithm does not provide us with the category of components (as opposed to precomponents). If we consider the space on the left, the outcome of the algorithm is shown in the middle, whereas the category of components is shown on the right:

6.3.3 The Seifert–Van Kampen Theorem

From a theoretical point of view, the category of components of a space can be computed from subspaces as follows. Suppose given a loop-free d-space X together with two subspaces Y and Z forming an open cover of X. In Sect. 4.31, we have shown that the fundamental category of X could be computed from those of Y and Z by a suitable pushout. Similarly, the category of components of X can be computed from those of Y and Z, as we now explain. The diagram on the left below, whose maps are the obvious inclusions, is a pushout in **dTop**:

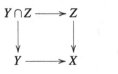
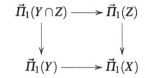

By Theorem 4.52, its image under the functor $\vec{\Pi}_1$, shown on the right above, is still a pushout in **Cat**. Suppose we have fixed a system Σ_Y (resp. Σ_Z) of weak isomorphisms in $\vec{\Pi}_1(Y)$ (resp. $\vec{\Pi}_1(Z)$), such that Σ_Y and Σ_Z are contained in the greatest system of weak isomorphisms of $\vec{\Pi}_1(X)$. As a consequence of Proposition 6.12, we can define Σ_X as the least system of weak isomorphisms containing both Σ_Y and Σ_Z. The inclusion maps then give rise to functors between the associated quotient categories:

Theorem 6.41 *The following diagram is a pushout in* **Cat***:*

$$
\begin{array}{ccc}
\vec{\Pi}_1(Y \cap Z) & \longrightarrow & \vec{\Pi}_1(Z)/\Sigma_Z \\
\downarrow & & \downarrow \\
\vec{\Pi}_1(Y)/\Sigma_Y & \longrightarrow & \vec{\Pi}_1(X)/\Sigma_X
\end{array}
$$

Example 6.42 Consider again the d-space X of Example 6.40, and write Y and Z for the two halves separated by the vertical dashed line in the middle:

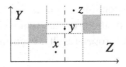

(technically, the spaces are a bit bigger than shown by the separation line so that they overlap). Notice that a dipath from x to y is in the greatest system of weak isomorphisms of $\vec{\Pi}_1(Y)$, but not in the one of $\vec{\Pi}_1(Z)$, which explains why we cannot suppose in general that Σ_Y and Σ_Z are greatest systems of weak isomorphisms in the above theorem. A suitable choice for Σ_Y and Σ_Z generates a set Σ_X which is the greatest system of weak isomorphisms of X and thus allows us to compute the category of components of the space, which was described in Example 6.40.

As explained in the example above, one of the drawbacks of this generalized form of Van Kampen theorem is that one essentially has to "guess" the component category in order to choose the right sets Σ_x, Σ_Y, and Σ_Z of weak isomorphisms, but the theorem still proves quite useful in practice. For instance, it allows the computation of the category of components of programs with conditional branchings: the syntax of the program defines in a quite straightforward manner the suitable covering of their geometric semantics with subspaces of the form handled in Sect. 6.1.

6.4 Historical Notes, Applications, and Extensions

The first steps concerning categories of components appeared in [52] and were later defined and studied thoroughly in Haucourt's PhD thesis [85], see also [68], where weak isomorphisms were called "Yoneda morphisms." These categories are not purely of theoretical interest. They provide a compact way of describing the fundamental category of simple programs, and thus to perform verification on those efficiently as explained in Chaps. 2 and 3: Theorem 6.23 ensures that we can use the presentation of the category of components of a program to build representatives for every execution traces up to dihomotopy, which is enough to cover all the possible behaviors of the program if we suppose that it is coherent, see Sect. 3.3.3.

Various extensions of the notion of component have also been investigated, with less success for now, and we report on those possible variations below.

6.4.1   Categories with Loops

So far, categories of components have been properly defined for loop-free categories only, i.e., in practice for the fundamental categories of programs without while loops. An obvious example which is not covered is the fundamental category of the directed circle $\vec{S}^1$ for which the fundamental category of components $\vec{\Pi}_0(\vec{S}^1)$ should clearly be the category corresponding to the additive monoid $\mathbb{N}$. Unfortunately, a direct generalization of most of the previous (equivalent) definitions of categories of components do not provide proper results, even in this simple case, as we now explain.

The morphisms of the fundamental category $\vec{\Pi}_1(\vec{S}^1)$ are in one-to-one correspondence with the elements $S^1 \times \mathbb{N} \times S^1$: a path $f : x \twoheadrightarrow y$ is characterized by the triple (x, n, y) consisting of its source x, its target y, and its winding number n. In order to guess what the greatest system of weak isomorphisms of $\vec{S}^1$ should be, we suppose that it is pure, and stable under both composition and the group of automorphisms of $\vec{S}^1$ (these assumptions are legitimate because they can be proven in the loop-free case). As a consequence, a morphism $(x, 0, y)$ is a weak isomorphism whenever (x, n, y) is so, for some $n \in \mathbb{N}$. Moreover, it readily follows from stability under automorphisms that if some morphism of the form $(x, 0, y)$ is a weak isomorphism, then so are all the others. As these morphisms generate the category $\vec{\Pi}_1(\vec{S}^1)$, one has to cope with the following dilemma: either $\vec{\Pi}_1(\vec{S}^1)$ has no weak isomorphism except identities, or every morphism is a weak isomorphisms. We favor the second case, since in the first one the category of components would be isomorphic to the original category. Writing Σ for the collection of morphisms of $\vec{\Pi}_1(\vec{S}^1)$, we observe that the localization $\vec{\Pi}_1(\vec{S}^1)[\Sigma^{-1}]$ is the fundamental groupoid of the circle, and that the quotient $\vec{\Pi}_1(\vec{S}^1)/\Sigma$ boils down to the terminal category. In particular, Theorem 6.27 is no longer available in the presence of loops which makes Definition 6.17 questionable. In fact it might be more reasonable to define the category of components as the full subcategory whose objects are in the image of a choice function (see Theorem 6.30). Indeed, we would then obtain $\vec{\Pi}_0(\vec{S}^1) = \mathbb{N}$ which seems more appropriate, even though it raises the technical problem of whether it is always possible to find a choice function which preserves the reachability of components.

The problem of the category of components of programs with loops can also be tackled through the universal covering space of their geometric semantics, which generally does not contain loops (see Sect. 4.3.2). In particular, this approach strengthens the belief that the category of components of the directed circle is the monoid $\mathbb{N}$.

6.4.2 Past and Future Components

The current definition of components identifies objects which intuitively have the same past and the same future up to dihomotopy, as formalized by the notion of weak isomorphism (Definition 6.4). By analogy with automata theory or bisimulation, one might also be interested in identifying points which only have the same past (resp. future) up to dihomotopy, giving rise to the notion of *past components* (resp. *future components*). Attempts to define those have been made [68, 70], but their theory still lacks many of the nice properties associated with the usual categories of components as presented in this chapter.

Chapter 7
Path Spaces

The space of dipaths in the geometric semantics of a program is generally very large, even for the most simple programs. In this chapter, we describe a method that allows one to "compress" this space in the case of simple programs, and to provide a finite combinatorial description of it which retains its essential topological characteristics: we compute a combinatorial model of this space, whose geometric realization is homotopy equivalent to it. This model will be a prod-simplicial complex (a variant of a presimplicial complex). Perhaps surprisingly at first glance, and in contrast to what happens for path spaces without the directedness assumption, this shows that those (functional) spaces have the homotopy type of a finite (CW-)complex: the space of directed paths itself is considered as a nondirected space.

The key to this description is a decomposition of a space of dipaths into subspaces of particular shapes, called restricted spaces, which are geometrically very simple: they are either empty or contractible. Interestingly, every restricted space can be coded by a certain boolean matrix, where inclusion of subspaces corresponds to the natural partial order on these matrices, and this gives rise to efficient computations. The algorithm that we present was originally introduced in [146] and further developed in [48]. To use it, one has to determine whether a particular restricted space is empty or not which can be done by adapting the deadlock algorithm described in Sect. 5.2.

In Sect. 7.1, we describe the part of the algorithm that is sufficient to compute the set of dipaths modulo dihomotopy, which is perhaps the most interesting from a verification point of view. In Sect. 7.2, we provide detailed proofs and extend the algorithm in order to construct a combinatorial model for the entire path space. Finally, we discuss some further extensions. We will mainly focus on simple programs, but will also mention how this can be extended to more general cases.

© Springer International Publishing Switzerland 2016

L. Fajstrup et al., *Directed Algebraic Topology and Concurrency*,

DOI 10.1007/978-3-319-15398-8_7

7.1 An Algorithm for Computing Components of Trace Spaces

7.1.1 Path Spaces for Simple Programs

Given two points x and y of a d-space X, we write $X(x, y)$ for the subspace of $X^{\vec{I}}$ consisting of all dipaths from x to y, equipped with the compact-open topology, and call it the *path space* from x to y. In particular, a *total path* is an element of $\check{G}_p(0, 1)$ (see Definition 4.13), i.e., a path from the beginning point $0 = (0, \ldots, 0)$ to the end point $1 = (1, \ldots 1)$, and we will be mostly interested in those paths in the following. The path components of $X(0, 1)$ correspond to the dihomotopy classes of total dipaths in X, and we shall call these *dipath classes* for short. In this section, we will concentrate on how to get hold on the set of dipath classes for a space X corresponding to a simple program: as explained in Sect. 3.3.3, these cover all the possible behaviors in a coherent program since two dihomotopic paths will lead to the same results, and can thus theoretically be used in order to verify a program.

Remark 7.1 Instead of considering dipaths, one may be interested in *traces*, which are paths modulo increasing reparametrizations, and the resulting "trace spaces" are often studied in the literature [41, 145]. Reparametrization equivalent dipaths are easily seen to be dihomotopic; moreover, path spaces and trace spaces are homotopy equivalent. For the sake of simplicity of the presentation, we focus here on path spaces.

As in the previous chapters, we restrict for simplicity to simple programs of given dimension n. We recall from Sect. 4.1.4 that the geometric semantics of such a program $p = p_1 \| \ldots \| p_n$ is of the form

$$X = \check{G}_p = \vec{I}^n \setminus \bigcup_{i=1}^{l} R^i \quad \text{with} \quad R^i = \prod_{j=1}^{n}]x_j^i, y_j^i[$$

with, for every $i \in [1 : l]$ and $j \in [1 : n]$, $x_j^i, y_j^i \in \{-\infty\} \cup I \cup \{\infty\}$ and $x_j^i < y_j^i$. We moreover suppose that all resources are of capacity $n - 1$ (this is only to ease the presentation, as explained in Remark 7.23): this amounts to supposing that all the x_j^i and y_j^i are different from $-\infty$ and ∞. For instance, we will consider the semantics of the following program p, which was already presented in Example 4.31:

$$p \quad = \quad P_a ; V_a ; P_b ; V_b \| P_b ; V_b ; P_a ; V_a$$

$$(7.1)$$

7.1.2 The Index Poset

In order to study the total dipaths in the path space $X(0, 1) = \check{G}_p(0, 1)$ of a state space X up to dihomotopy, we focus on how they "turn around" each of the holes in X. Here is a quick instructive description in dimension 2. For a given hole (for instance the hole denoted 1 in (7.2) below), a dipath has either to stay below that hole or to the left of it. Only one of these options occurs if the hole is extended in parallel to one of the axes: extending a hole like the dark gray hole 1 parallel to the vertical axis (i.e., carving the light gray "below hole 1" in the first situation) to the boundary of the state space forces every dipath to stay to the left of that hole; extending it parallel to the horizontal axis (light gray "to the left of 1" like in the second situation) forces dipaths to stay below hole 1.

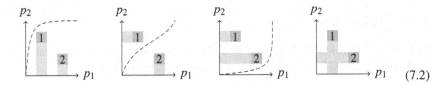

$$(7.2)$$

Consider all possible combinations of extensions like the four ones depicted in (7.2). In some situations (the three first ones in our example) there exists a total dipath with the given combination of behaviors; in others, the occurrence of a new deadlock (as in the last case) shows that there is no total dipath with the described combination of behaviors.

It turns out that this combinatorial information is enough to compute the space of directed paths up to homotopy equivalence. The path space from (7.2) with the two original dark gray holes denoted 1 and 2 has in fact three contractible components.

A simple way to encode the combinatorial information about the extension of holes and of resulting dipaths is through boolean matrices.

Notation 7.2 We write $\mathcal{M}_{l,n}$ for the poset of $l \times n$ *matrices* $M = (M(i,j))$, with l rows and n columns and coefficients in $\mathbb{Z}_2$ (the boolean field of two elements). This set of matrices is given a partial order via the entrywise ordering $0 \leqslant 1$, i.e., $M \leqslant N$ whenever $M(i,j) \leqslant N(i,j)$ for every pair of indices $(i,j) \in [1:l] \times [1:n]$. For later reference, we write $\mathcal{M}_{l,n}^R$ (resp. $\mathcal{M}_{l,n}^C$) for the subposet consisting of matrices such that each *row* contains *at least one* coefficient equal to 1 (resp. each *column* contains *exactly one* coefficient equal to 1).

Definition 7.3 Given a matrix $M \in \mathcal{M}_{l,n}$, we write $\mathcal{R}(M) \subseteq X \subseteq \vec{I}^n$ for the *restricted* subspace induced by M, obtained by extending downward each forbidden rectangle R^i in every direction $j' \neq j$ for every j such that $M(i,j) = 1$. Formally,

$$\mathcal{R}(M) = \vec{I}^n \setminus \bigcup_{M(i,j)=1} \tilde{R}_j^i \tag{7.3}$$

where

$$\tilde{R}^i_j = \prod_{j'=1}^{j-1}[0, y^i_{j'}[\times]x^i_j, y^i_j[\times \prod_{j'=j+1}^{n} [0, y^i_{j'}[$$

The spaces $\tilde{R}^i_j$ are sometimes informally referred to as *walls* because of their geometric representation, see Examples 7.6 and 7.7 below. It is immediate from the definitions that $M \leqslant N$ implies $\mathscr{R}(N) \subseteq \mathscr{R}(M)$. In other words, given $X = \check{G}_p$, we may conclude the following:

Lemma 7.4 *The operation $\mathscr{R}$ can be considered as a functor from the opposite of the posetal category $(\mathscr{M}_{l,n}, \leqslant)$ to* **dTop**.

In order to study whether there is a total path in the space associated to a matrix, we introduce the following notions.

Definition 7.5 A matrix M is **dead** if $\mathscr{R}(M)(0, 1) = \emptyset$, i.e., there is no total dipath in the d-space $\mathscr{R}(M)$, and **alive** otherwise. In particular, we write

$$\mathscr{M}_{\text{dead}} = \{M \in \mathscr{M}_{l,n} \mid M \text{ is dead}\}$$

and

$$\mathscr{M}_{\text{alive}} = \{M \in \mathscr{M}^{\text{R}}_{l,n} \mid M \text{ is alive}\}$$

The **index poset** that we will work with is the set $\mathscr{M}_{\text{alive}}$ equipped with the partial order described in Notation 7.2.

In the definition of $\mathscr{M}_{\text{alive}} \subseteq \mathscr{M}^{\text{R}}_{l,n}$, alive matrices are supposed to have a coefficient 1 in each row, i.e., each hole should be extended in at least one direction. This will make sure that the space $\mathscr{R}(M)$ is geometrically very simple for a matrix $M \in \mathscr{M}_{\text{alive}}$ as shown in Proposition 7.9.

Example 7.6 In the introductory example, the three extensions of holes (7.2) are, respectively, encoded by the following matrices:

$$\begin{pmatrix} 1 & 0 \\ 1 & 0 \end{pmatrix} \qquad \begin{pmatrix} 0 & 1 \\ 1 & 0 \end{pmatrix} \qquad \begin{pmatrix} 0 & 1 \\ 0 & 1 \end{pmatrix} \qquad \begin{pmatrix} 1 & 0 \\ 0 & 1 \end{pmatrix}$$

The last matrix is dead: this indicates that there is no dipath passing to the left of hole 1 and below hole 2. The three other matrices are alive.

Example 7.7 The geometric semantics of the program consisting of three copies of the thread $P_a ; V_a ; P_b ; V_b$ in parallel, with $\kappa_a = \kappa_b = 2$, is

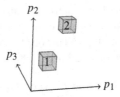

The spaces $\mathscr{R}(M)$ corresponding to the matrices

$$\begin{pmatrix} 1 & 0 & 0 \\ 0 & 0 & 1 \end{pmatrix} \qquad \begin{pmatrix} 0 & 0 & 1 \\ 1 & 0 & 0 \end{pmatrix} \qquad \begin{pmatrix} 0 & 0 & 0 \\ 1 & 1 & 1 \end{pmatrix}$$

are respectively

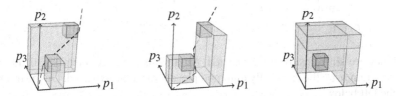

The first two matrices are alive, as shown by the drawn total paths, and the last one is dead.

From Lemma 7.4, we may conclude:

Lemma 7.8 *The set $\mathscr{M}_{\text{dead}}$ is upward closed within $\mathscr{M}_{l,n}$, and the set $\mathscr{M}_{\text{alive}}$ is downward closed within $\mathscr{M}_{l,n}^{R}$.*

The set $\mathscr{M}_{\text{dead}}$ (resp. $\mathscr{M}_{\text{alive}}$) is thus completely characterized by its minimal (resp. maximal) elements.

The matrices in the index poset are suitable objects in the study of dipath classes because the associated spaces $\mathscr{R}(M)$ are topologically very simple, as formalized in the following proposition that will be proved in Proposition 7.26.

Proposition 7.9 *For any matrix $M \in \mathscr{M}_{l,n}^{R}$, the space $\mathscr{R}(M)(x, y)$ of dipaths from x to y is either empty or contractible. Hence, a matrix in $\mathscr{M}_{l,n}^{R}$ is alive (resp. dead) if and only if $\mathscr{R}(M)(0, 1)$ is contractible (resp. empty).*

In particular, if M is alive, then any two dipaths in $\mathscr{R}(M)(x, y)$ are dihomotopic. Moreover, we will now explain that the dipath classes can be recovered from an equivalence relation on the matrices in $\mathscr{M}_{\text{alive}}$:

Definition 7.10 Two matrices $M, N \in \mathscr{M}_{l,n}^{R}$ are *connected* if they are related by the smallest equivalence relation containing $\leqslant$ on $\mathscr{M}_{l,n}^{R}$.

In particular, if the maximum of two matrices is alive, then the two matrices are connected. Intuitively, alive matrices describe sets of mutually dihomotopic total paths. Whenever the maximum of two matrices is alive, there are paths which satisfy the constraints imposed by both matrices, i.e., all dipaths satisfying the constraints of either matrix are mutually dihomotopic. This observation is reflected in the following result that we prove in Propositions 7.27 and 7.28.

Proposition 7.11 *There is a bijection between the set of connected components of $\mathcal{M}_{\text{alive}}$ and the set of dipath classes in X.*

Example 7.12 Consider the "floating cube" program $p = P_a ; V_a \| P_a ; V_a \| P_a ; V_a$. The associated path space X_p is a cube from which an interior cube has been removed. The matrices in $\mathcal{M}_{\text{alive}}$ are, along with the associated restricted spaces,

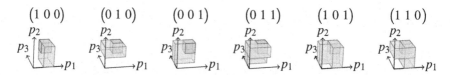

and they are all (transitively) connected. All dipaths in the geometric realization of the program p are thus mutually dihomotopic, as was to be expected. On the triangle on the left below

points correspond to minimal matrices in $\mathcal{M}_{\text{alive}}$, and edges to maximal such matrices. In fact, we will see that this hollow triangle provides a model for the path space. The interior of the triangle is not present since the matrix $M = \begin{pmatrix} 1 & 1 & 1 \end{pmatrix}$ corresponds to an empty path space $\mathcal{R}(M)(0, 1)$: there are too many walls for a dipath to get through, as shown on the right.

7.1.3 Determination of Dipath Classes

The computation of the dihomotopy classes of total paths (dipath classes for short) in the geometric semantics $X = \check{G}_p$ of a given simple program p will be performed in three steps.

1. We compute the set $\mathcal{M}_{\text{dead}}$ of dead matrices; by Lemma 7.8, it is actually enough to determine the minimal ones in $\mathcal{M}_{\text{dead}}^{\min} = \mathcal{M}_{\text{dead}} \cap \mathcal{M}_{l,n}^{C}$ with exactly one entry equal to 1 in each column.

2. We use $\mathcal{M}_{\text{dead}}^{\text{min}}$ to compute the index poset $\mathcal{M}_{\text{alive}}$; it is enough to determine the maximal ones in $\mathcal{M}_{\text{alive}}^{\text{max}}$.
3. We deduce the set of dipath classes by determining the quotient of $\mathcal{M}_{\text{alive}}$ with respect to connectedness.

In order to determine which matrices are dead, we start with the following simple observation: when a matrix $M \in \mathcal{M}_{l,n}$ is dead, there is no total path in $\mathcal{R}(M)$ $(0,1)$. In this case, any maximal dipath starting at 0 ends at a point $x < 1$, and x is thus a deadlock. If this is true for every maximal dipath, then 0 is in the doomed region (see Definition 4.44) for one of the deadlocks that hence has to be produced by one of the walls (see Definition 7.3) for each of the n directions, i.e., the matrix M has at least one entry equal to 1 in each column. On the other hand, if n walls (from a matrix $M \in \mathcal{M}_{l,n}^{\text{C}}$ with exactly one entry one per column) result in a deadlock, then the associated doomed region has the start point 0 as its minimal vertex, see Definition 7.3 and Algorithm 5.14. Hence, no dipath starting at 0 can avoid this deadlock. We have thus shown:

Lemma 7.13 *A matrix $M \in \mathcal{M}_{l,n}^{\text{C}}$ is dead if and only if the space $\mathcal{R}(M)$ contains a deadlock with 0 in the doomed region.*

We apply the characterization of deadlocks in geometric semantics given in Theorem 5.11 of Sect. 5.2 to find the dead matrices by checking a number of inequalities. This requires introducing further notation: given a subset I of $[1:l]$ and an index $j \in [1:n]$, we write $y_j^I = \min\left\{y_j^i \mid i \in I\right\}$, where $y_j^\emptyset = \infty$ by convention. Given a matrix $M \in \mathcal{M}_{l,n}$, we define the set of *nonzero rows* of M by $R(M) = \{i \in [1:l] \mid \exists j \in [1:n], M(i,j) \neq 0\}$. For a matrix $M \in \mathcal{M}_{l,n}^{\text{C}}$, we write $i : [1:n] \to [1:l]$ for the function characterized by $M(i(j),j) = 1$.

Proposition 7.14 *A matrix $M \in \mathcal{M}_{l,n}^{\text{C}}$ is dead iff it satisfies $x_j^{i(j)} < y_j^{R(M)}$ for every $j \in [1:n]$.*

Proof Following Theorem 5.11 a deadlock with 0 in the doomed region can only be formed by the n walls $\tilde{R}_j^{i(j)}$. Those intersect in the region $\prod_j]x_j^{i(j)}, y_j^{R(M)}[$, which is nonempty if and only if the condition holds. $\square$

Example 7.15 In the example below with $l = 2$ and $n = 2$, the matrix $M = \begin{pmatrix} 0 & 1 \\ 1 & 0 \end{pmatrix}$ is dead:

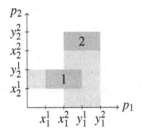

$$x_2^1 = 1 < 2 = y_2^{\{1,2\}}$$
$$x_1^2 = 2 < 3 = y_1^{\{1,2\}}$$

Example 7.16 Consider the geometric semantics of the second program of Example 4.31. The minimal dead matrices are

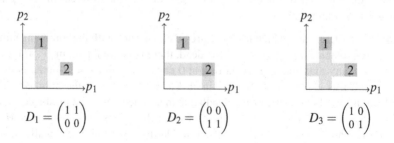

$$D_1 = \begin{pmatrix} 1 & 1 \\ 0 & 0 \end{pmatrix} \qquad D_2 = \begin{pmatrix} 0 & 0 \\ 1 & 1 \end{pmatrix} \qquad D_3 = \begin{pmatrix} 1 & 0 \\ 0 & 1 \end{pmatrix}$$

The above proposition enables us to compute the set of (minimal) dead matrices, for instance by enumerating all matrices in $\mathcal{M}_{l,n}^C$ and checking whether they satisfy condition of Proposition 7.14 (a more efficient method is described in Sect. 7.1.4). From this set, the index poset $\mathcal{M}_{\text{alive}}$ can be determined using the following property:

Lemma 7.17 *A matrix $M \in \mathcal{M}_{l,n}$ is alive if and only if, for every matrix $N \in \mathcal{M}_{\text{dead}}$, we have $N \not\leqslant M$, i.e., there exists indices $i \in [1 : l]$ and $j \in [1 : n]$ such that $M(i, j) = 0$ and $N(i, j) = 1$.*

Since the poset $\mathcal{M}_{\text{alive}} \subseteq \mathcal{M}_{l,n}^R$ is downward closed by Lemma 7.8, it is sufficient to determine the subset $\mathcal{M}_{\text{alive}}^{\max}(X)$ of maximal matrices. Lemma 7.17 provides a simple-minded algorithm to achieve that purpose.

Algorithm 7.18 We write $\mathcal{M}_{\text{dead}} = \{D_1, \ldots, D_p\}$. Then we compute the sets C_k of maximal matrices $M \in \mathcal{M}_{l,n}^R$ such that $D_i \not\leqslant M$ for every $i \in [1 : k]$. We start from the set $C_0 = \{\mathbf{1}\}$ where $\mathbf{1}$ is the matrix containing only 1 as coefficients. Given a matrix M, we write $M^{\neg(i,j)}$ for the matrix obtained from M by replacing the (i, j)th coefficient by $1 - M(i, j)$. The set C_{k+1} is then computed from C_k by iteratively performing the following steps for all matrices $M \in C_k$ such that $D_k \leqslant M$:

1. Remove M from C_k,
2. For every (i, j) such that $D_k(i, j) = 1$, and hence $M(i, j) = 1$, if there exists no matrix $N \in C_k$ such that $M^{\neg(i,j)} \leqslant N$ and if $M^{\neg(i,j)} \in \mathcal{M}_{l,n}^R$, add $M^{\neg(i,j)}$ to C_k.

The set $\mathcal{M}_{\text{alive}}^{\max}$ is obtained as C_p.

Remark 7.19 In the previous algorithm, if we replace the second point by

2. for every (i, j) such that $D_k(i, j) = 1$ and $M^{\neg(i,j)} \in \mathcal{M}_{l,n}^R$, add $M^{\neg(i,j)}$ to C_k.

we compute a set C_p such that $\mathcal{M}_{\text{alive}}^{\max} \subseteq C_p \subseteq \mathcal{M}_{\text{alive}}$, which is enough to compute connected components, and is faster to compute in practice. Other implementations of the algorithm can be obtained by reformulating the computation of $\mathcal{M}_{\text{alive}}^{\max}$ as finding a minimal transversal in a hypergraph, for which efficient algorithms have been proposed [98].

Example 7.20 Consider again Example 4.31. The algorithm starts with

$$C_0 = \left\{ M_0 = \begin{pmatrix} 1 & 1 \\ 1 & 1 \end{pmatrix} \right\}$$

For C_1, in order to achieve $D_1 \not\leqslant M_0^{\neg(1,j)}$, we change any of the two ones in the first row into a zero:

$$C_1 = \left\{ M_1 = \begin{pmatrix} 0 & 1 \\ 1 & 1 \end{pmatrix}, M_2 = \begin{pmatrix} 1 & 0 \\ 1 & 1 \end{pmatrix} \right\}$$

Similarly for C_2, we have to change the bits on the second row so that $D_2 \not\leqslant M_r^{\neg(2,j)}$:

$$C_2 = \left\{ M_3 = \begin{pmatrix} 0 & 1 \\ 0 & 1 \end{pmatrix}, M_4 = \begin{pmatrix} 0 & 1 \\ 1 & 0 \end{pmatrix}, M_5 = \begin{pmatrix} 1 & 0 \\ 0 & 1 \end{pmatrix}, M_6 = \begin{pmatrix} 1 & 0 \\ 1 & 0 \end{pmatrix} \right\}$$

Finally, we have $D_3 \not\leqslant M_r$ for $r = 3, 4, 6$. We have to exclude M_5 since $D_3 \leqslant M_5$ and $M_5^{\neg(i,j)} \notin \mathscr{M}_{2,2}^{R}$ for $(i,j) = (1,0)$ and $(i,j) = (0,1)$. We conclude that $\mathscr{M}_{\text{alive}} = \mathscr{M}_{\text{alive}}^{\max} = C_3 = \{M_3, M_4, M_6\}$. The path spaces corresponding to those matrices are the three first depicted in (7.2). Since no two different of those matrices are connected in $\mathscr{M}_{\text{alive}}$, there are exactly three dipath classes (one class in each of the restricted spaces).

Remark 7.21 For $n = 2$, no alive matrix $M \in \mathscr{M}_{1,2}^{R}$ is connected to another alive matrix $N \in \mathscr{M}_{1,2}^{R}$. Every row is either $(0\ 1)$ or $(1\ 0)$. A matrix $N \leqslant M$ with $N \neq M$ has a row (00) and is therefore not contained in $\mathscr{M}_{1,2}^{R}$. A matrix $N \geqslant M$ with $N \neq M$ has a row $(1\ 1)$ that causes an empty path space $\mathscr{R}(N)(0,1)$ and hence $N \in \mathscr{M}_{\text{dead}}$. As a consequence, for $n = 2$, there is a bijection between dipath classes and alive matrices.

Remark 7.22 In order to determine dipath classes in X, one has to determine whether two maximal alive matrices M and N are connected. To do this in practice, it helps to use the following characterization: two matrices M and N are connected if their meet $M \wedge N \in \mathscr{M}_{1,n}$ (taken coordinatewise) belongs to $\mathscr{M}_{1,n}^{R}$. Take the transitive closure of the relation defined by that condition. This coincides with the relation "connected" from Definition 7.10.

Remark 7.23 We have supposed up to now that resources were of capacity $n - 1$, which amounts to imposing that the forbidden region is a product of intervals of the form $]x_j^i, y_j^i[$ where x_j^i and y_j^i are different from $-\infty$ and ∞. This requirement can be dropped, i.e., we can also handle cases where some of the forbidden regions intersect the boundaries of $\vec{I}^n$ in some directions.

We write $B \in \mathscr{M}_{1,n}$ for the *boundary matrix* with entries $B(i,j) = 0$ whenever $x_j^i = -\infty$ (i.e., the ith hyperrectangle touches the lower boundary of $\vec{I}^n$ in dimension j) and $B(i,j) = 1$ otherwise. The matrices in $\mathscr{M}_{\text{dead}}$ are then the matrices $M \in \mathscr{M}_{1,n}$ of the form $M = N \wedge B$, for some matrix $N \in \mathscr{M}_{1,n}^{C}$, which satisfy the condition of

Proposition 7.14 and such that for every $j \in C(M)$, we have $y_j^{R(M)} = \infty$, where $C(M)$ is the set of indices of zero columns of M. Based on similar techniques, all subsequent developments can also be adapted to this case [48, 147].

7.1.4   An Efficient Implementation

Of course, it is not a good idea to check the condition from Proposition 7.14 on all l^n matrices $M \in \mathcal{M}_{l,n}^C$ in order to compute the set $\mathcal{M}_{\mathrm{dead}}$ of dead matrices. Instead, we go through these matrices columnwise from left to right and eliminate candidates "as soon as possible". The condition may fail after considering only the first few columns, and then it will fail for every matrix with these first columns: a subset of the columns gives rise to a submatrix M' with a subset $R(M') \subseteq R(M)$ of nonzero rows. If $x_j^i \geqslant y_j^{R(M')}$ for one of the nonzero entries (i, j) coefficients, the matrix M cannot satisfy the condition of Proposition 7.14 because $x_j^i \geqslant y_j^{R(M')} \geqslant y_j^{R(M)}$.

The actual function computing the dead matrices is presented in Fig. 7.1, in pseudo-OCaml code. This recursive function investigates the jth column of a matrix M (whose columns with index less than j are supposed to be handled already)

```
let rec compute_dead j m rows yrows =
  if j = n then dead := m :: !dead else
    for i = 0 to l - 1 do
      try
        let changed_rows = not (Set.mem i rows) in
        let rows = Set.add i rows in
        let m = Array.copy m in
        (match m.(j) with
          | Some i → if x_j^i ⩾ yrows.(j) then raise Exit
          | None → if yrows.(j) ≠∞ then raise Exit);
        let yrows =
          let j' = j in
          if not changed_rows then yrows else
            Array.mapi (fun j yrj →
              if yrj ⩽ y_j^i then yrj else
                match m.(j) with
                | None →
                    if j ⩽j' && y_j^i≠∞ then raise Exit; y_j^i
                | Some i →
                    if x_j^i ⩾ y_j^i then raise Exit; y_j^i
              ) yrows
        in
        compute_dead (j+1) m rows yrows
      with Exit → ()
    done
```

Fig. 7.1 Algorithm for computing dead matrices

and performs the check: it tries to set $M(i, j)$ to 1 (and all the others to 0) for every index $i \in [0 : l - 1]$ (in this code, indices are starting from 0 instead of 1 as customary in programming languages). If a matrix with these first j columns is alive, the computation is aborted by raising the Exit exception. Only when all n columns pass the test, the complete matrix is added to the list *dead* of dead matrices. Since a matrix $M \in \mathcal{M}_{l,n}^C$ has exactly one nonzero coefficient in a given column, it will be coded as an array of length n whose jth element is either None when all the elements of the jth column are null, or Some i when the ith coefficient of the jth column is 1 and the others are 0. The argument *rows* is the set of indices of known nonzero rows of M and *yrows* is an array of length n such that $yrows.(j) = y_j^{rows}$. Note that the algorithm takes advantage of the fact that when the coefficient i chosen for the jth column is already in *rows* (i.e., when the variable *changed_rows* is false) then many computations can be saved because the coefficients y_j^{rows} are not changed.

Once the set of dead matrices is determined, the set $\mathcal{M}_{\text{alive}}$ of alive matrices is computed using the naive algorithm of Sect. 7.1.3, as explained in Example 7.20. Finally, the representatives of paths are computed as the connected components (in the sense of Proposition 7.11) of $\mathcal{M}_{\text{alive}}$, in a straightforward way. An explicit sequence of instructions corresponding to every representative M can easily be computed: it corresponds to the sequence of instructions crossed by any increasing total path in the d-space $\mathcal{R}(M)$.

7.2 Combinatorial Models for Path Spaces

In this section, we prove some of the properties stated in Sect. 7.1. Using more advanced tools from algebraic topology, we follow up and identify a path space $X(x, y)$ with a simplicial complex that allows one to reason about—and sometimes to compute—higher topological invariants of the path spaces of interest, such as its homology groups. In particular, this often allows one to qualitatively distinguish path spaces with the same number of path components (i.e., dihomotopy classes of dipaths).

7.2.1 Contractibility of Restricted Path Spaces

We begin by showing that the restricted spaces $\mathcal{R}(M)$ are either empty or contractible, i.e., that they are homotopy equivalent to a point. Recall from Sect. 4.2.4 that the space $\vec{I}^n$ can be equipped with the product order, i.e., $x \leqslant y$ whenever $x_i \leqslant y_i$ for every $i \in [1 : n]$, and that the resulting poset forms a lattice with the join $x \vee y$ (resp. meet $x \wedge y$) taken as the componentwise max (resp. min) of x and y.

A d-space X described by a cubical region is not stable under joins in general. For instance, consider the space $X = \vec{I}^2 \setminus]\frac{1}{3}, \frac{2}{3}[^2$. The points $x = (\frac{1}{2}, 0)$ and $y = (0, \frac{1}{2})$

are contained in X, but their join $x \vee y = (\frac{1}{2}, \frac{1}{2})$ is not. The following lemma however shows that the restricted spaces $\mathcal{R}(M)$ are closed under joins:

Lemma 7.24 *For an arbitrary matrix $M \in \mathcal{M}_{l,n}^{R}$ and points $x, y \in \mathcal{R}(M)$, the sub-space $\mathcal{R}(M) \cap [x, y]$ is closed under joins.*

Proof The intersection of spaces that each are closed under joins is closed under joins itself. A hypercube $[x, y]^n$ is clearly closed under joins. Since we have $\mathcal{R}(M) = \bigcap_{M(i,j)=1} \vec{I}^n \setminus \tilde{R}_j^i$, it is enough to show that $\vec{I}^n \setminus \tilde{R}_j^i$ is stable under joins for any given (i, j). This is easily done by inspection. $\square$

This operation allows us to construct dihomotopies between paths in such a space as follows. Given a space Z, the *endpoint map* $e : Z^I \to Z \times Z$ is the function which to a path $f : I \to Z$ associates $e(f) = (f(0), f(1))$.

Proposition 7.25 *For $M \in \mathcal{M}_{l,n}^{R}$, any two paths $f, g \in \mathcal{R}(M)(x, y)$ with same source x and target y are dihomotopic. Moreover, there is a continuous section h of the end point map $e : \mathcal{R}(M)(x, y)^I \to \mathcal{R}(M)(x, y) \times \mathcal{R}(M)(x, y)$.*

Proof The continuous function $I \times \vec{I} \to \mathcal{R}(M)(x, y)$ defined by $(s, t) \mapsto f(t) \vee g(st)$ is a dihomotopy between f and $f \vee g$, which is well-defined by Lemma 7.24 and contained in $\mathcal{R}(M)(x, y)^I$. Similarly, we can define a dihomotopy from $f \vee g$ to g, and by concatenating the two, we obtain a dihomotopy $h(f, g) : I \to \mathcal{R}(M)(x, y)$ from f to g. The function h thus defined is continuous in f and g. $\square$

As a consequence, path spaces $\mathcal{R}(M)(x, y)$ are topologically very simple:

Proposition 7.26 *For every matrix $M \in \mathcal{M}_{l,n}^{R}$ and every pair of points $x, y \in \mathcal{R}(M)$, the path space $\mathcal{R}(M)(x, y)$ is either empty or contractible.*

Proof Suppose that $\mathcal{R}(M)(x, y)$ is nonempty and choose an arbitrary path f that it contains. We can define a homotopy $H : I \times \mathcal{R}(M)(x, y) \to \mathcal{R}(M)(x, y)$ by $H(t, g) = h(f, g)(t)$, where h is the map constructed in Proposition 7.25. It contracts the space $\mathcal{R}(M)(x, y)$ to the element f since we have $H(0, g) = f$ for every $g \in \mathcal{R}(M)(x, y)$. $\square$

The above proposition allows us to show Proposition 7.11 from Sect. 7.1.2 via the following two steps:

Proposition 7.27 *For every dipath $f \in X(0, 1)$ there exists a matrix $M \in \mathcal{M}_{l,n}^{R}$ such that $f \in \mathcal{R}(M)(0, 1)$.*

Proof It is clear that $X_1(0, 1) \cap X_2(0, 1) = (X_1 \cap X_2)(0, 1)$ for subspaces $X_1, X_2 \subseteq \vec{I}^n$. It is thus enough to consider the statement in the case where $X = \vec{I}^n \setminus R$ with one removed hyperrectangle $R = \prod_{j=1}^{n}]x_j, y_j[$ corresponding to a matrix with one nonzero row. For any dipath $f \in X(0, 1)$, consider $t_1 = \min \{ t \mid \exists j \in [1 : n], f_j(t) = y_j \}$, and t_0 such that $t_0 < t_1$ and $t_0 < t < t_1$ implies $x_j < f_j(t) < y_j$, for some index j such that $f_j(t_1) = y_j$. Since $f(t) \notin R$ for all such t, there exists $i \neq j$ such that $f_i(t) \leqslant x_i$ for $t_0 < t < t_1$. Since the path f is directed, we have that $0 \leqslant t \leqslant t_1 \Rightarrow f_i(t) \leqslant x_i$ and $t_1 \leqslant t \leqslant 1 \Rightarrow f_j(t) \geqslant y_j$, hence $f \in \mathcal{R}(M_i)(0,1)$ with M_i the row matrix with a single entry 1 in column i. $\square$

Proposition 7.28 *Suppose given two alive matrices* $M, N \in \mathcal{M}_{\text{alive}}$. *Two dipaths* $f \in \mathcal{R}(M)(0, 1)$ *and* $g \in \mathcal{R}(N)(0, 1)$ *are dihomotopic if and only if* M *and* N *are connected.*

Proof For the right-to-left implication, it is enough to consider the case where $M \leqslant N$ (the case where $N \leqslant M$ is similar, and the general case is obtained by transitivity): since $\mathcal{R}(N) \subseteq \mathcal{R}(M)$ and both spaces are contractible, the result follows. We now consider the left-to-right implication. The space of all alive matrices $\mathcal{M}_{\text{alive}}$ decomposes into equivalence classes with respect to the connectedness equivalence relation: $\mathcal{M}_{\text{alive}} = \bigsqcup_J \mathcal{M}_J$. The subspaces $\mathcal{R}(\mathcal{M}_J)(0, 1) = \bigcup_{M_i \in \mathcal{M}_J} \mathcal{R}(M_i)(0, 1)$ are both open and closed. Let $H \in X(0, 1)^I$ denote a dihomotopy between $f \in \mathcal{R}(M)(0, 1)$ and $g \in \mathcal{R}(N)(0, 1)$. By Proposition 7.27, intermediate paths $H(t)$ for $t \in I$, are contained in $\mathcal{R}(M_t)(0, 1)$ for appropriate (alive) matrices $M_t \in \mathcal{M}_{l,n}^R$ and hence in $\mathcal{R}(M_J)(0, 1)$ for exactly one J. For each J, the set of all $t \in I$ such that $H^{-1}(\mathcal{R}(M_J)) = \{t \in I \mid H_t \in \mathcal{R}(M_J)\}$ is open, and the $H^{-1}(\mathcal{R}(M_J))$ form a cover of the interval I by disjoint open sets. Hence the entire interval is contained in $H^{-1}(\mathcal{R}(M_J))$ for a single J; in particular, M and N are connected. $\qquad\square$

7.2.2 Presimplicial Sets and the Nerve Theorem

Recall from Sect. 3.4 that a precubical set consists of sets of abstract n-cubes together with face relations. A presimplicial set is the analog where n-cubes are replaced by n-triangles, which are more generally called n-simplices [84].

Definition 7.29 A *presimplicial set* (or *semi-simplicial set* or *Δ-set*) D consists of a family $(D_n)_{n \in \mathbb{N}}$ of sets, whose elements are called *n-simplices*, together with maps $\partial_{n,i} : D_n \to D_{n-1}$, for $n \geqslant 1$ and $0 \leqslant i \leqslant n$, such that

$$\partial_{n-1,i} \circ \partial_{n,j} = \partial_{n-1,j-1} \circ \partial_{n,i}$$

for $0 \leqslant i < j \leqslant n$. We often write ∂_i instead of $\partial_{n,i}$ in the following. The category of presimplicial sets can also be described as the presheaf category $\hat{\Delta}$, where Δ is the *presimplicial category* whose objects are integers and a morphism $f : m \to n$ is an injective increasing function $f : [0 : m] \to [0 : n]$.

Example 7.30 For instance the presimplicial set D, which can be pictured as

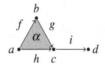

is defined by $D_0 = \{a, b, c, d\}, D_1 = \{f, g, h, i\}, D_2 = \{\alpha\}$ and $D_n = \emptyset$ for $n > 2$, and faces are given by $\partial_1(f) = \partial_1(h) = a$, $\partial_0(f) = \partial_1(g) = b$, $\partial_0(g) = \partial_0(h) = \partial_1(i) = c$, $\partial_0(i) = d$, $\partial_0(\alpha) = g$, $\partial_1(\alpha) = h$ and $\partial_2(\alpha) = f$.

As in the case of precubical sets, we have an adjunction between nerve and geometric realization functors, see Sect. 3.4.2.

Definition 7.31 Given a cocomplete category $\mathscr{C}$ and a functor $I : \Delta \to \mathscr{C}$, there are two induced functors:

- the *nerve* $N_I : \mathscr{C} \to \hat{\Delta}$ defined on $A \in \mathscr{C}$ by $N_I(A) = \mathscr{C}(I-, A)$,
- the *realization* $R_I : \hat{\Delta} \to \mathscr{C}$ defined on $D \in \hat{\Delta}$ by $R_I(D) = \mathrm{colim}(y/D \xrightarrow{\pi} \Delta \xrightarrow{I} \mathscr{C})$.

Proposition 7.32 *The realization functor is left adjoint to the nerve functor.*

We will be mostly interested in the following particular instances of these functors. Firstly, consider the functor $\Delta : \Delta \to \mathbf{Top}$, sending n to the subspace of $\mathbb{R}^n$, called *standard n-simplex* and denoted Δ^n, such that $\Delta^n = \{(x_1, \ldots, x_n) \in \mathbb{R}^n_+ \mid \sum_{i=1}^n x_i = 1\}$. The associated realization functor can be described as follows.

Definition 7.33 The *geometric realization* functor $|-| : \hat{\Delta} \to \mathbf{Top}$ associates to each presimplicial set D the topological space $|D| = \coprod_{n \in \mathbb{N}} (D_n \times \Delta^n) / \approx$, where $\approx$ is an equivalence relation identifying a point on the border of a simplex with the corresponding point in its border simplex. A topological space obtained as the geometric realization of a presimplicial set is called a *simplicial complex*.

Secondly, consider the functor $I : \Delta \to \mathbf{Cat}$ such that $I(n)$ is the category associated to the poset $[0 : n]$ equipped with the usual total order, called the *categorical n-simplex*. The associated nerve functor $N_I : \mathbf{Cat} \to \hat{\Delta}$ is often simply denoted N.

It is well known that the notion of colimit is not compatible with homotopy equivalences: if we take a diagram in **Top** and replace one of the spaces by a homotopy equivalent space, the colimits of the original and of the modified diagram will generally not be homotopy equivalent to each other. For instance, consider the pushout on the left below, where both maps are inclusion from the discrete space with two points as the two endpoints of the a segment: the pushout has the homotopy type of a 1-sphere. However, if we replace the two segments by points (which are homotopy equivalent), as shown on the right, we obtain a point as pushout, which is not equivalent to a 1-sphere.

(7.4)

This is why we will consider the following variant of the notion of a colimit, called a *homotopy colimit*, which can be thought of as the best way to correct the notion of a colimit so that it respects homotopy equivalences. To build an intuition about it, consider a pushout diagram

$$X \xleftarrow{\ f\ } Z \xrightarrow{\ g\ } Y \tag{7.5}$$

The construction of its colimit starts from the disjoint union $X \sqcup Y$, and quotients it by the equivalence relation $\approx$ such that $f(z) \approx g(z)$ for every $z \in Z$. The homotopy pushout will also start from $X \sqcup Y$, but instead of quotienting it, it will "add a line" (i.e., a copy of Δ^1) from $f(z)$ to $g(z)$ for every $z \in Z$.

Example 7.34 The homotopy colimits of the two diagrams (7.4) are the unions of the mapping cylinders of the two maps over the two-point base; they can be drawn as

Both homotopy colimits are homotopy equivalent to the 1-sphere S^1.

The general definition of homotopy colimits is a generalization of this idea: starting from the disjoint union of spaces, it adds lines for relations between points, triangles for relations between relations, and so on. We describe it only very briefly, after introducing some preliminary definitions, because a detailed presentation is unfortunately out of the scope of this book, see [17, 36] for instance. In this paragraph only, we write Δ for the simplicial category, which is defined as the presimplicial category (Definition 7.29) except that morphisms are increasing functions (not required to be injective), and call *simplicial space* a functor $\Delta^{\mathrm{op}} \to \mathbf{Top}$. Given a diagram $F : J \to \mathbf{Top}$ of topological spaces, its *simplicial replacement* is the simplicial space $\mathrm{srep}\, F : \Delta^{\mathrm{op}} \to \mathbf{Top}$ such that, given $n \in \Delta^{\mathrm{op}}$, we have

$$\mathrm{srep}\, Fn \quad = \quad \coprod_{j_0 \leftarrow j_1 \leftarrow \ldots \leftarrow j_n} F(j_n)$$

where the coproduct is indexed by chains of n composable morphisms in J. The simplicial replacement can thus be seen as a diagram in **Top** of the form

$$\mathrm{srep}\, F \quad = \quad \coprod_{j_0} F(j_0) \rightrightarrows \coprod_{j_0 \leftarrow j_1} F(j_1) \underset{\Lbag}{\overset{\Rbag}{\rightrightarrows}} \coprod_{j_0 \leftarrow j_1 \leftarrow j_2} F(j_2) \qquad \ldots$$

where the morphisms are face maps $\partial_i : \mathrm{srep}\, F(n+1) \to \mathrm{srep}\, F(n)$, for $0 \leqslant i \leqslant n$, and degeneracy maps $\sigma_i : \mathrm{srep}\, F(n) \to \mathrm{srep}\, F(n+1)$, for $0 \leqslant i < n$, which are defined as follows. The face map ∂_i, for $0 \leqslant i < n$, maps $F(j_{n+1})$ at $j_0 \leftarrow j_1 \leftarrow \ldots \leftarrow j_{n+1}$ by the identity map to itself at $j_0 \leftarrow \ldots \leftarrow j_i \circ j_{i+1} \leftarrow \ldots \leftarrow j_{n+1}$, whereas ∂_n maps $F(j_{n+1})$ at

$j_0 \leftarrow j_1 \leftarrow \ldots \leftarrow j_{n+1}$ by $F(j_n \leftarrow j_{n+1})$ to $F(j_n)$ at $j_0 \leftarrow j_1 \leftarrow \ldots \leftarrow j_n$. Degeneracies can be described similarly by adding extra identities to chains of composable morphisms. Finally, given a simplicial space $X : \Delta^{\mathrm{op}} \to \textbf{Top}$, its *geometric realization* $|X|$ is the following coequalizer

$$|X| \quad = \quad \mathrm{coeq}\left(\coprod_{f:m\to n} Xn \times \Delta^m \underset{Xn \times \Delta^f}{\overset{Xf \times \Delta^m}{\rightrightarrows}} \coprod_k Xk \times \Delta^k \right)$$

Note that we recover the usual geometric realization of a simplicial set when each Xn is a discrete space.

Definition 7.35 The *homotopy colimit* hocolim F over a diagram $F : J \to \textbf{Top}$ of topological spaces is the space $|\mathrm{srep}\, F|$.

Example 7.36 The homotopy colimit of a diagram corresponding to a single map $f : X \to Y$ (over the diagram $\cdot \to \cdot$) is the mapping cylinder of the map f, i.e., the space $((X \times [0, 1]) \sqcup Y)/\approx$, where $\approx$ identifies $(x, 1)$ with $f(x)$ for each $x \in X$.

Example 7.37 Given a pushout diagram (7.5), its homotopy colimit is the space $((X \sqcup Z \sqcup Y) \times \Delta^0 \sqcup (Z \sqcup Z) \times \Delta^1)/\approx$, where $\approx$ identifies points as in the figure below:

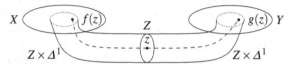

It can easily be checked that we recover the computations of Example 7.34.

The homotopy lemma below [17, 102, 151, 164] allows for the construction of homotopy equivalences between homotopy colimits by constructing them "levelwise" on the diagrams. A morphism $\phi : F \Rightarrow G$ between two diagrams $F, G : J \to \textbf{Top}$ of the same shape J is a natural transformation between the functors: it consists in a natural collection of continuous maps $\phi_j : Fj \to Gj$ indexed by objects $j \in J$. Such a morphism induces a map hocolim ϕ : hocolim $F \to$ hocolim G between homotopy colimit, which satisfies the following.

Theorem 7.38 (Homotopy lemma) *If for every object j of J the map ϕ_j is a homotopy equivalence, then* hocolim ϕ : hocolim $F \to$ hocolim G *is a homotopy equivalence as well.*

We are particularly interested in the case where diagrams of spaces arise from an open cover $U = (U_i)_{i\in I}$ of a topological space X with I a totally ordered set. Such a cover induces a poset PU consisting of all *nonempty* finite subsets $K = \{i_0 < \cdots < i_n\} \subseteq I$ such that the space $U_K = \bigcap_{i\in K} U_i$ is nonempty, ordered by reverse inclusion. We regard PU as a category (associated to the poset), and also as a presimplicial set whose n-simplices are the sets K of cardinality $n+1$ as above, the maps $\partial_{n,k}$ removing i_k. The *Čech nerve* of the covering U is the geometric realization $|PU|$ of the associated presimplicial set.

Example 7.39 Consider the following space consisting of contractible subspaces of the plane, shown in gray on the left; these subspaces have two-by-two contractible intersections but the intersection of all three is empty. The Čech nerve of the covering of their union is thus a hollow triangle, as pictured on the right: each vertex corresponds to one of the subspaces, each edge to one of the nonempty intersections.

This easy example illustrates that a space which is covered by several contractible subspaces whose intersections are either empty or contractible is homotopy equivalent to the nerve of the covering.

The topology of the space X can, under the conditions of the Nerve Theorem 7.42 below, be partially recovered as follows. We consider two functors S and T from PU (regarded as a category) to **Top**, defined on an object $K \in PU$ by

- $S(K) = U_K$, and reverse inclusion $K \supseteq K'$ corresponds to inclusion $U_K \subseteq U_{K'}$,
- $*(K) = *$, the one-point space.

Remark 7.40 Observe that the space $X = \bigcup_{i \in I} U_i$ can be recovered as the colimit of S and that the homotopy colimit of $*$ is, by definition, the Čech nerve:

$$X = \operatorname{colim} S \qquad\qquad |PU| = \operatorname{hocolim} *$$

On the other hand, for a path-connected set X, $\operatorname{colim} *$ is just a point.

To compare homotopy colimits and colimits, one may associate to any diagram of spaces $F : J \to$ **Top** a fiber projection map $p : \operatorname{hocolim} F \to \operatorname{colim} F$ forgetting the second coordinates (in the simplices Δ^n). It enjoys good properties if J comes from the category PU associated to a covering of a paracompact topological space [17, 102, 151, 164]: A topological space is called *paracompact* if it is Hausdorff and if every open covering admits a locally finite refinement. Note that every compact space, every metric space, every CW-complex (and thus every simplicial or cubical complex) is paracompact.

Theorem 7.41 (Projection lemma) *Given a paracompact topological space X and a finite open cover $U = (U_i)_{i \in I}$, the projection* $\operatorname{hocolim} S \to \operatorname{colim} S$ *is a homotopy equivalence.*

The nerve theorem below is attributed to Borsuk [15], the presentation here follows Kozlov [102].

Theorem 7.42 (Nerve theorem) *Suppose given a paracompact topological space X and a finite open cover $U = (U_i)_{i \in I}$ such that every intersection U_K, with $\emptyset \neq K \subseteq I$, is either empty or contractible. Then the Čech nerve $|PU|$ of the covering U is homotopy equivalent to X.*

Proof We compare the two spaces X and $|PU|$ with hocolim S. The natural transformation $\phi : S \to *$ sending each (nonempty) contractible intersection U_K to $*$ induces a homotopy equivalence on homotopy colimits by Theorem 7.38. Applying Theorem 7.41, we end up with a homotopy equivalence

$$X \;=\; \operatorname{colim} S \;\leftarrow\; \operatorname{hocolim} S \;\to\; \operatorname{hocolim} * \;=\; |PU|$$

which allows us to conclude.   □

7.2.3 Path Space as a Prod-Simplicial Complex

We will show that the space of all dipaths $X(0, 1)$ from 0 to 1 in the space $X = \check{G}_p$ admits a particularly simple and tractable algebraic description in the form of a variant of a simplicial complex: a *prod-simplicial complex*, which is a topological space obtained by gluing products of simplices (i.e., spaces of the form $\Delta^{i_1} \times \cdots \times \Delta^{i_l}$), see [102]. It turns out that the path space $X(0, 1)$ is homotopy equivalent to such a (finite-dimensional) prod-simplicial complex. To give an idea, we look at a few easy examples; the results will be explained at the end of this section.

Example 7.43 In dimension two (two threads in parallel), the associated complex is concentrated in dimension 0 and thus finite. Hence, the path space $X(0, 1)$ is homotopy equivalent to the disjoint union of finitely many contractible (and therefore connected) path components.

Example 7.44 We have already looked at the "floating cube" from Example 7.12 in dimension 3, where the corresponding complex is a hollow triangle. In any dimension $n \geqslant 2$, the "floating hypercube" arises from a program that runs $P_a ; V_a$ in parallel n times, where a is a resource of capacity $n - 1$. The corresponding complex is a hollow $(n - 1)$-simplex homotopy equivalent to a sphere S^{n-2}.

Example 7.45 A program running $P_a ; V_a ; P_b ; V_b$ in parallel n times, where a and b are resources of capacity $n - 1$, results in a forbidden region consisting of two floating hypercubes on the "diagonal", see Example 7.7 for the case $n = 3$. It leads to a path space $X(0, 1)$ homotopy equivalent to a product $S^{n-2} \times S^{n-2}$.

We first need to take a closer look at the posetal categories of binary matrices introduced in Sect. 7.1.2.

Definition 7.46 A matrix $M \in \mathcal{M}_{l,n}^{\mathrm{R}}$ is *basic* if every row vector is a unit vector, i.e., with a single entry equal to one. We write $\mathcal{M}_{l,n}^{\mathrm{R}^*}$ for the set of such matrices.

The set $\mathcal{M}_{l,n}^{R^*}$ generates the lattice $\mathcal{M}_{l,n}^{R}$ (under $\vee$, the componentwise maximum). The minimal alive matrices form a subset $\mathcal{M}_{\text{alive}}^{\min} = \mathcal{M}_{\text{alive}} \cap \mathcal{M}_{l,n}^{R^*}$. One checks easily that $\mathcal{R}(M) \cap \mathcal{R}(N) = \mathcal{R}(M \vee N)$ and

$$\mathcal{R}(M)(0, 1) \cap \mathcal{R}(N)(0, 1) = \mathcal{R}(M \vee N)(0, 1) \tag{7.6}$$

Note that even when M and N are alive, $M \vee N$ may be dead. The path space $X(0, 1)$ is covered by the contractible subspaces of the form $\mathcal{R}(M)(0, 1)$ with $M \in \mathcal{M}_{\text{alive}}^{\min}$. Multiple intersections of such contractible path spaces correspond to the spaces $\mathcal{R}(M)(0, 1)$ with $M \in \mathcal{M}_{l,n}^{R}$ that can be dead or alive. More formally,

Definition 7.47 We write $\mathcal{P} : \mathcal{M}_{\text{alive}}^{\text{op}} \to \textbf{Top}$ for the *total path functor* which to a matrix M associates the space of total paths $\mathcal{P}M = \mathcal{R}(M)(0, 1)$. A relation $M \leqslant N$ induces an inclusion of path spaces $\mathcal{P}N \subseteq \mathcal{P}M$.

As an immediate consequence of Proposition 7.27 and of (7.6), we conclude:

Lemma 7.48 $X(0, 1) = \text{colim } \mathcal{P}$.

Remark 7.49 It is in general not true that the space X itself is a colimit of subspaces $\mathcal{R}(M)(0, 1)$ corresponding to alive matrices. For instance, no deadlock is contained in any of these subspaces.

Remark 7.50 The spaces $\mathcal{P}M$ are generally not open, but one may replace them by subspaces that are a little larger and still contractible [146].

As a consequence, the path space $X(0, 1)$ is homotopy equivalent to the nerve of the covering given by the spaces $\mathcal{P}M$ with $M \in \mathcal{M}_{\text{alive}}^{\min}$. This representation has the disadvantage that the same subspace $\mathcal{P}M$ can occur as intersection of many different choices of subspaces represented by basic matrices: the same matrix in $\mathcal{M}_{\text{alive}}$ may arise as the least upper bound of many choices of basic matrices, e.g.

$$\begin{pmatrix} 1 & 1 & 0 \\ 1 & 1 & 0 \end{pmatrix} = \begin{pmatrix} 1 & 0 & 0 \\ 1 & 0 & 0 \end{pmatrix} \vee \begin{pmatrix} 0 & 1 & 0 \\ 0 & 1 & 0 \end{pmatrix} = \begin{pmatrix} 1 & 0 & 0 \\ 0 & 1 & 0 \end{pmatrix} \vee \begin{pmatrix} 0 & 1 & 0 \\ 1 & 0 & 0 \end{pmatrix}$$

To avoid the resulting combinatorial complexity, we apply a more conceptual way to represent path space $X(0, 1)$ that exploits the fact that the data can be viewed as glued from *products* of simplices, each simplex corresponding to a forbidden hyperrectangle R^i. More precisely, we consider the following functor:

Definition 7.51 We define the *geometric realization* of a nonzero binary vector $B \in \{0, 1\}^n$ as the face $\mathcal{T}B$ of the $(n-1)$-simplex Δ^{n-1} consisting of points which are spanned by the unit vectors e_j such that $B(j) = 1$. It consists of the points $(x_1, \ldots, x_n) \in \mathbb{R}^n$ satisfying $\sum_{j=1}^{n} x_j = 1$ and $B(j) = 0$ implies $x_j = 0$. The *geometric realization* of a matrix $M \in \mathcal{M}_{l,n}$ is then defined as the product of the geometric realizations of its rows: $\mathcal{T}M = \prod_{i=1}^{l} \mathcal{T}M(i)$, where $M(i)$ denotes the ith row of M. This operation defines a functor $\mathcal{T} : \mathcal{M}_{\text{alive}} \to \textbf{Top}$. An inequality $M \leqslant N$ corresponds to the canonical inclusion of spaces $\mathcal{T}M \hookrightarrow \mathcal{T}N$.

Example 7.52 The matrices M and N below have the following geometric realizations:

$$\mathscr{T}M = \mathscr{T}\begin{pmatrix} 1 & 1 & 0 \\ 1 & 1 & 0 \end{pmatrix} = \boxed{} \qquad \mathscr{T}N = \mathscr{T}\begin{pmatrix} 1 & 1 & 1 \\ 1 & 1 & 0 \end{pmatrix} = \boxed{}$$

Note that $M \leqslant N$ results in an embedding of $\mathscr{T}M$ into $\mathscr{T}N$.

If we take the colimit of the spaces $\mathscr{T}M$ over the matrices M in the posetal category $\mathscr{M}_{\text{alive}}$, i.e., along the functor $\mathscr{T}$, we obtain a prod-simplicial complex:

Definition 7.53 We write $\mathbf{T}X = \text{colim } \mathscr{T}$.

In the following, we compare the colimits of the functors $\mathscr{P}$ and $\mathscr{T}$ to their homotopy colimits (see Definition 7.35).

Theorem 7.54 *The path space $X(0, 1)$ is homotopy equivalent to the prod-simplicial complex* $\mathbf{T}X$.

Proof In analogy with the proof of the nerve Theorem 7.42, we compare both colimit spaces $X(0, 1)$ and $\mathbf{T}X$ with the *homotopy* colimit of the trivial functor $* : \mathscr{M}_{\text{alive}} \to \mathbf{Top}$ sending each object into one-point space $*$. Note that, by definition of homotopy colimits, we have $\text{hocolim} * = |N(\mathscr{M}_{\text{alive}})| = \text{hocolim}_{\mathscr{M}_{\text{alive}}^{op}} *^{op}$. For every matrix $M \in \mathscr{M}_{\text{alive}}$, the (unique) continuous map $\mathscr{P}M \to *$ (resp. $\mathbf{T}(M) \to *$) is a homotopy equivalence, combining to give a natural transformation $\phi : \mathscr{P} \Rightarrow *$ (resp. $\psi : \mathscr{T} \Rightarrow *^{op}$). By Theorem 7.38, the induced maps on homotopy colimits

$$\text{hocolim } \phi : \text{hocolim } \mathscr{P} \to |N(\mathscr{M}_{\text{alive}})| \quad \text{and} \quad \text{hocolim } \psi : \text{hocolim } \mathscr{T} \to |N(\mathscr{M}_{\text{alive}})|$$

are homotopy equivalences as well. Moreover, by Theorem 7.41, the fiber projection maps $\text{hocolim } \mathscr{P} \to \text{colim } \mathscr{P} = X(0, 1)$ and $\text{hocolim } \mathscr{T} \to \text{colim } \mathscr{T} = \mathbf{T}(X)$ are homotopy equivalences. To sum up, we have constructed the following sequence of homotopy equivalences

$$\text{colim } \mathscr{P} \leftarrow \text{hocolim } \mathscr{P} \to \text{hocolim } *^{op} \leftrightarrow \text{hocolim } * \leftarrow \text{hocolim } \mathscr{T} \to \text{colim } \mathscr{T}$$

which are connecting $X(0, 1)$ (on the left) with $\mathbf{T}X$ (on the right). $\qquad \square$

Let us go through the cases announced in Examples 7.43, 7.44 and 7.45.

Example 7.55 Consider again Example 7.43. In dimension 2, the poset $\mathscr{M}_{\text{alive}}$ consists of matrices with rows either $\begin{pmatrix} 1 & 0 \end{pmatrix}$ or $\begin{pmatrix} 0 & 1 \end{pmatrix}$. Hence, the partial order is trivial and $\mathbf{T}X$ is a discrete space with as many points as there are matrices in $\mathscr{M}_{\text{alive}}$. Theorem 7.54 tells us that $X(0, 1)$ consists of contractible components, one for each alive matrix.

Example 7.56 Consider again Example 7.44. For a forbidden region consisting of a single floating hypercube in dimension n, the associated matrices in $\mathcal{M}_{\text{alive}}$ are all single binary row vectors apart from $(0 \ldots 0)$ and $(1 \ldots 1)$. The associated prodsimplicial complex $\mathbf{TX} \simeq X(0, 1)$ is the hollow $(n-1)$-simplex $\partial \Delta^{n-1} \simeq S^{n-2}$. It is hollow since at least one coordinate has to be equal to 0.

Example 7.57 Consider again Example 7.45. For a forbidden region consisting of l floating hypercubes in dimension n along the diagonal, the alive matrices have row vectors as in the previous example. Hence, the associated prod-simplicial complex $\mathbf{TX} \simeq X(0, 1)$ is a product $\prod_1^l \partial \Delta^{n-1} \simeq \prod_1^l S^{n-2}$, whose homology is concentrated in dimensions $i \times (n-2)$ with $0 \leqslant i \leqslant l$.

Example 7.58 A forbidden region consisting of two or more hyperrectangles all intersecting in a hyperrectangle R gives rise to a path space $X(0, 1)$ homotopy equivalent to a single sphere S^{n-2}. In fact, this path space is homotopy equivalent to the path space $(\vec{I}^n \setminus R)(0, 1)$.

Example 7.59 For a forbidden region consisting of hyperrectangle obstructions that are not totally orderable, the associated complex $\mathbf{TX}$ is often far more complicated. We only mention one example here. Let $F = R_1 \cup R_2$ denote a forbidden region consisting of two disjoint incomparable hyperrectangles $R_1, R_2 \subseteq \vec{I}^n$ and let $Y = \vec{I}^n \setminus F$. For $n = 3$, such a state space corresponds, for example, to the program

$$\mathsf{P}_a ; \mathsf{V}_a ; \mathsf{P}_b ; \mathsf{V}_b \parallel \mathsf{P}_a ; \mathsf{V}_a ; \mathsf{P}_b ; \mathsf{V}_b \parallel \mathsf{P}_b ; \mathsf{V}_b ; \mathsf{P}_a ; \mathsf{V}_a$$

where a and b have capacity 2. For $n > 2$, the space $Y(0, 1)$ is homotopy equivalent to the one-point union $S^{n-2} \vee S^{n-2}$, e.g. with $n = 3$:

In particular, it is connected and its homology is concentrated in dimensions 0 and $n - 2$. For $n = 2$, it consists of three contractible connected components.

More generally, the homology groups of path spaces $\check{G}_p(0, 1)$ corresponding to parallel programs consisting of n strings each of the form $\mathsf{P}_{a_1} ; \mathsf{V}_{a_1} ; \ldots ; \mathsf{P}_{a_k} ; \mathsf{V}_{a_k}$ (every lock is relinquished before a new one is acquired; the resources a_i all have capacity $n - 1$ but need not all be different) have been determined [149]: their homology is concentrated in dimensions that are integer multiples of $n - 2$.

Remark 7.60 Far more general spaces can be obtained as path spaces. It has been shown [172] that for any finite-dimensional simplicial complex P on n vertices, there exists a simple program p with n threads, hence with a geometric semantics $X = \check{G}_p \subseteq \vec{I}^n$, such that $X(0, 1)$ is homotopy equivalent to the disjoint union of the complex P and a sphere S^{n-2}. For instance, a triangulation of the real projective plane $P\mathbb{R}^2$ on 6 vertices (and 10 triangles) gives rise to a space X with trace space $X(0, 1)$ which is homotopy equivalent to $P\mathbb{R}^2 \sqcup S^4$.

The previous methods and results can be extended to path spaces of more general spaces, not necessarily generated by simple programs. In particular, programs with loops are handled in [48]: the general idea is that the extension of holes can only induce a finite number of spaces in each of the loops, and the trace space can thus be described by some sort of automaton called the *shadow automaton*, see [47]. More generally, an analysis of path spaces in a general precubical complex (see Sect. 3.4) has been performed in [148]. Current work [123] attempts to replace the prod-simplicial complex **T**X by constructions making use of configuration space techniques. In most cases, the dimension of the resulting complex decreases dramatically. There are even benefits for the determination of components corresponding to dihomotopy classes of dipaths.

Chapter 8
Perspectives

We hope that this panorama of relationships between directed algebraic topology and concurrency has given the reader an impression of the profound links between the two fields. Even though most computational processes are intrinsically discrete, adopting a geometric point of view to this investigation has brought many new insights, and we believe that this is only a beginning. This book can only serve as a relatively short introduction to the subject. Many related results and approaches were not presented, and we would like to point the reader to some of those in this final chapter.

Concurrency Control in Distributed Databases

A distributed database can be seen as a shared-memory machine on which processes, often called *transactions*, act by reading and writing, getting permissions to do so by using the appropriate functions on attached semaphores. One of the main concerns in this area is to maintain the coherence of the distributed database while ensuring good performance. This is achieved through a definition of suitable policies (or *protocols*) for transactions to perform their own actions, by using P and V operations. Of course, in this context, the deadlock-freedom of transactions is of great importance. The correctness of a distributed database is itself very often expressed by some form of a *serializability* or *linearizability* condition. Testing serializability is unfortunately known to be a NP-complete problem [137], even when the model is only based on simple binary semaphores.

A geometric approach to the study of distributed databases was initiated in [169] and continued in [138], both relying on a directed topological approach to the semantics of databases. Further work made this more explicit, with a notion of directed homotopy, as in [118] (see also the last chapter of [142]) linked with the serializability condition. Improvements of algorithms were given later in [169] and in [155], among others. For deadlock detection, the directed topological approach was put forward in [50]. An application to proving the serializability of 2-phase locked protocols was given in [82], and in [51], explicitly using directed homotopy.

© Springer International Publishing Switzerland 2016
L. Fajstrup et al., *Directed Algebraic Topology and Concurrency*,
DOI 10.1007/978-3-319-15398-8_8

Fault-Tolerant Protocols for Distributed Systems

A natural extension to the work on databases is to consider the problem of committing values, in case of faults (either in the underlying network or of computing nodes). This is one of the core subjects of the field of fault-tolerant protocols [119]. The seminal result in this field was established by Fisher et al. in 1985 [55]. They proved that there exists a simple task that cannot be solved in a message-passing (or equivalently a shared memory) system with at most one potential crash. In particular, there is no way in such a distributed system to solve the very fundamental consensus problem: each processor starts with an initial value in local memory (typically an integer); they all are to end up with a common value, which is one of the initial values. This created a very active research area, see for instance [88, 119]. Later on, Biran, Moran, and Zaks developed a characterization of the decision tasks that can be solved by a (simple) message-passing system in the presence of one failure [11]. The argument uses a "similarity chain," which could be seen as a connectedness result of a representation of the space of all reachable states, also called the *view complex* [103] or the (full-information) *protocol complex* [90].

This argument turned out to be difficult to extend to models with more failures, as higher-connectedness properties of the view complex matter in these cases. This technical difficulty was first tackled, using homological calculations, by Herlihy and Shavit [89] (and independently [14, 150]): there are simple decision tasks, such as consensus once again that cannot be solved in the wait-free asynchronous model, i.e., shared-memory distributed protocols on n processors, with up to $n - 1$ crash failures. The full characterization of wait-free asynchronous decision tasks with atomic read and write on registers was described by Herlihy and Shavit [90]. Their analysis relies on the central notion of chromatic (or colored) simplicial complexes, and subdivisions of those. All results above are deduced from the contractibility of the so-called "standard" chromatic subdivision, which was completely formalized in [103]. It corresponds to the *view complex* of distributed algorithms solving layered immediate snapshot protocols. The central fact that the (iterated) protocol complexes in that case are contractible (in fact, even collapsible) was shown independently by Kozlov [104] and some of the authors of this book [74] in an effort to link this (classical) topological approach, to the directed topological approach, started in [64, 65] and partially solved in [75]. We refer the reader to the excellent book [88] for further details.

Higher-Dimensional Automata and Semantic Equivalences

The combinatorial approach to bisimulation equivalences between cubical models was only hinted at in Sect. 3.5. It is very interesting and illuminating to make the comparison between Higher-Dimensional Automata (HDA), as introduced originally in [141, 160] and further studied in [63], and other classical models for concurrency. Apart from transition systems, one may consider (prime) event structures, asynchronous transition systems, (safe) Petri nets, etc. Classical adjunctions between the latter are fully described in [167], and adjunctions with HDA are developed in [73]. HDAs are the most expressive model for concurrency currently in use; they even generalize unsafe or "general" Petri nets [161]. To be complete in such comparisons, we would have had to introduce semantic equivalences, such as bisimulation.

For geometric models such as HDA, some form of bisimulation equivalence has been studied since the very beginning [25, 160], and then fully developed in terms of open maps [96], by generalizing history preserving bisimulations [39] and homotopy preserving bisimulations [40]. Very early on, it was suggested that these semantic equivalences could be studied using some topological invariants: the idea was that some geometric obstruction should explain why two models are not bisimilar [72]. Hence some authors developed homology theories for directed topology, starting with [63, 72]. Unfortunately, a form of directed homology is very difficult to approach in a classical, abelian setting, as done in further work like [38, 77, 97]. Non-abelian approaches have been put forward originally by Krishnan, and developed by Dubut [35], in very promising approaches. A current line of research is to link directed homology with some aspects of persistent homology [172] used for topological data analysis [23, 135].

Rewriting Techniques

Another instance of topological reasoning that seems related to directed topology is Squier's theorem in rewriting systems theory [156]. This theorem gives a necessary condition for the existence of a presentation of a given monoid by a finite canonical rewriting system in terms of its homology (it must be of finite dimension). It is definitely a computability result, in the same way as in fault-tolerant distributed systems theory, but for something which looks sequential (rewriting). As hinted in [63], it can be understood as a problem of concurrency theory since the study of confluence of rewriting systems is related to parallel reduction techniques (as in [114] for instance). The resolutions used in most of the proofs of this theorem [3, 54, 80, 81, 100, 107, 108] are very much like a Knuth–Bendix completion procedure, where higher-dimensional objects fill in possible defects of local confluence. This looks like building higher-dimensional transitions implementing the parallel (confluent) reductions (see in particular [80] where the resolution is given as a cubical complex; in dimension one it is generated by the transition system coming from the reduction relation). Some other proof techniques are reminiscent of directed homotopy, as in [157] for instance. Other interesting relations should be studied concerning "higher-dimensional" word problems, as in [22].

Homotopy Type Theory

Last but not least, homotopy type theory [163] has current extensions toward directed homotopy type theory [116, 117] that could lead to exciting new developments in this young but promising branch of mathematics. This setting provides a very natural framework to manipulate higher-dimensional groupoids, and one might hope that a directed variant could help manipulating higher-dimensional categories. Also, many recent models of homotopy type theory are based on (variants of) cubical sets [10, 115], which could constitute an interesting starting point in order to bridge the two fields.

References

1. F.E. Allen, Control flow analysis, in *ACM SIGPLAN Notices - Proceedings of a Symposium on Compiler Optimization, Number 7 in 5*, Association for Computing Machinery, July 1970, pp. 1–19
2. B. Alpern, F.B. Schneider, Recognizing safety and liveness. Distrib. Comput. **2**(3), 117–126 (1987)
3. D.J. Anick, On the homology of associative algebras. Trans. Am. Math. Soc. **296**, 641–659 (1986)
4. M. Arkowitz, *Introduction to Homotopy Theory*, Universitext (Springer, New York, 2011)
5. A. Arnold, *Systèmes de transitions finis et sémantique des processus communicants*, Études et recherches en informatique (Masson, 1992)
6. T. Balabonski, E. Haucourt, A geometric approach to the problem of unique decomposition of processes, *CONCUR 2010-Concurrency Theory* (Springer, Heidelberg, 2010), pp. 132–146
7. M.A. Bednarczyk, Categories of asynchronous systems. Ph.D. thesis, University of Sussex (1987)
8. M.A. Bednarczyk, A.M. Borzyszkowski, W. Pawlowski, Generalized congruences - epimorphisms in cat. Theory Appl. Categ. **5**(11), 266–280 (1999)
9. P.A. Bernstein, V. Hadzilacos, N. Goodman, *Concurrency Control and Recovery in Database Systems* (Addison-Wesley Longman Publishing Co., Inc., Boston, 1986)
10. M. Bezem, T. Coquand, S. Huber, A model of type theory in cubical sets, in *19th International Conference on Types for Proofs and Programs (TYPES 2013), Leibniz International Proceedings in Informatics (LIPIcs)*, vol. 26, Schloss Dagstuhl-Leibniz-Zentrum fuer Informatik (2014), pp. 107–128
11. O. Biran, S. Moran, S. Zaks, A combinatorial characterization of the distributed tasks which are solvable in the presence of one faulty processor, in *Proceedings of the Seventh Annual ACM Symposium on Principles of Distributed Computing* (ACM, 1988), pp. 263–275
12. R. Bonichon, G. Canet, L. Correnson, É. Goubault, E. Haucourt, M. Hirschowitz, S. Labbé, S. Mimram, Rigorous evidence of freedom from concurrency faults in industrial control software, in *SAFECOMP* (2011), pp. 85–98
13. F. Borceux, *Handbook of Categorical Algebra, I. Basic Category Theory, Encyclopedia of Mathematics and its Applications*, vol. 50 (Cambridge University Press, Cambridge, 1994)
14. E. Borowsky, E. Gafni, Generalized FLP impossibility result for t-resilient asynchronous computations, in *Proceedings of the 25th STOC* (ACM Press, 1993)
15. K. Borsuk, On the imbedding of systems of compacta in simplicial complexes. Fundamenta Mathematicae **35**, 217–234 (1948)
16. O. Bournez, O. Maler, A. Pnueli, Orthogonal polyhedra: representation and computation. Lect. Notes Comput. Sci. **1569**, 46–60 (1999)

© Springer International Publishing Switzerland 2016

L. Fajstrup et al., *Directed Algebraic Topology and Concurrency*,

DOI 10.1007/978-3-319-15398-8

17. A.K. Bousfield, D.M. Kan, *Homotopy Limits, Completions and Localizations*, vol. 304, Lecture Notes in Mathematics (Springer, New York, 1972)

18. J. Brazas, Generalized covering space theorems. Technical report, Georgia State University (2015)

19. R. Brown, *Topology and Groupoids* (BookSurge Publishing, North Charleston, 2006)

20. R. Brown, P.J. Higgins, On the algebra of cubes. J. Pure Appl. Algebr. **21**, 233–260 (1981)

21. P. Bubenik, K. Worytkiewicz, A model category for local pospaces. Homol., Homotopy Appl. **8**(1), 263–292 (2006)

22. A. Burroni, Higher-dimensional word problems with applications to equational logic. Theor. Comput. Sci. **115**(1), 43–62 (1993)

23. G. Carlsson, Topology and data. Bull. Am. Math. Soc. **46**, 255–308 (2009)

24. S.D. Carson, P.F.J. Reynolds, The geometry of semaphore programs. ACM Trans. Program. Lang. Syst. **9**(1), 25–53 (1987)

25. G.L. Cattani, V. Sassone, Higher dimensional transition systems, in *Proceedings of Logic in Computer Science Conference (LICS'96)* (1996), pp. 55–62

26. E.M. Clarke, My 27-year quest to overcome the state explosion problem, in *24th Annual IEEE Symposium on Logic in Computer Science, LICS'09* (2009), pp. 3–3

27. E.M. Clarke, E.A. Emerson, A.P. Sistla, Automatic verification of finite-state concurrent systems using temporal logic specifications. ACM Trans. Program. Lang. Syst. **8**(2), 244–263 (1986)

28. E.G. Coffman, M.J. Elphick, A. Shoshani, System deadlocks. Comput. Surv. **3**(2), 67–78 (1971)

29. P. Cousot, R. Cousot, Comparing the Galois connection and widening/narrowing approaches to abstract interpretation, in *Proceedings of the International Workshop Programming Language Implementation and Logic Programming, PLILP*, Invited paper (1992), pp. 269–295

30. V. Diekert, G. Rozenberg, *The Book of Traces* (World Scientific, Singapore, 1995)

31. E.W. Dijkstra, Cooperating sequential processes, in *Programming Languages: NATO Advanced Study Institute*, ed. by F. Genuys (Academic Press, New York, 1968), pp. 43–112

32. E.W. Dijkstra, The structure of the operating system. Commun. ACM **11**(15), 341–436 (1968)

33. A.B. Downey, *The Little Book of SEMAPHORES: The Ins and Outs of Concurrency Control and Common Mistakes*, 2nd edn. (Createspace Independent Publication, 2009)

34. J. Dreier, C. Ene, P. Lafourcade, Y. Lakhnech, On unique decomposition of processes in the applied π-calculus, in *FOSSACS 2013: 16th International Conference on Foundations of Software Science and Computation Structures* (2013)

35. J. Dubut, E. Goubault, J. Goubault-Larrecq, Natural homology, in *Automata, Languages, and Programming - 42nd International Colloquium, ICALP 2015, Proceedings, Part II* (Kyoto, Japan, 2015), 6–10 July, pp. 171–183

36. D. Dugger, A primer on homotopy colimits. Preprint (2008)

37. S. Eilenberg, Ordered topological spaces. Am. J. Math. **63**, 39–45 (1941)

38. U. Fahrenberg, Directed homology. Electron. Notes Theor. Comput. Sci. **100**, 111–125 (2004)

39. U. Fahrenberg, A. Legay, History-preserving bisimilarity for higher-dimensional automata via open maps. Electron. Notes Theor. Comput. Sci. **298**, 165–178 (2013)

40. U. Fahrenberg, A. Legay, Homotopy bisimilarity for higher-dimensional automata (2014). arXiv preprint arXiv:1409.5865

41. U. Fahrenberg, M. Raussen, Reparametrizations of continuous paths. J. Homotopy Relat. Struct. **2**(2), 93–117 (2007)

42. L. Fajstrup. Loops, ditopology and deadlocks. Math. Struct. Comput. Sci. **10**(4), 459–480 (2000) (Geometry and concurrency)

43. L. Fajstrup, Dicovering spaces. Homol. Homotopy Appl. **5**(2), 1–17 (2003)

44. L. Fajstrup, Dipaths and dihomotopies in a cubical complex. Adv. Appl. Math. **35**(2), 188–206 (2005)

45. L. Fajstrup, Classification of dicoverings. Topol. Appl. **157**(15), 2402–2412 (2010)

46. L. Fajstrup, Erratum to dicovering spaces. Homol. Homotopy Appl. **13**(1), 403–406 (2011)

47. L. Fajstrup, Trace spaces of directed tori with rectangular holes. Math. Struct. Comput. Sci. **24**(02), e240202 (2014)
48. L. Fajstrup, E. Goubault, E. Haucourt, S. Mimram, M. Raussen, Trace spaces: an efficient new technique for state-space reduction, *Programming Languages and Systems* (Springer, New York, 2012), pp. 274–294
49. L. Fajstrup, É. Goubault, E. Haucourt, M. Raussen, Components of the fundamental category. Appl. Categ. Struct. **12**(1), 81–108 (2004)
50. L. Fajstrup, É. Goubault, M. Raussen, Detecting deadlocks in concurrent systems, in *Proceedings of the 9th International Conference on Concurrency Theory* (Springer, New York, 1998)
51. L. Fajstrup, M. Raussen, É. Goubault, Algebraic topology and concurrency. Theor. Comput. Sci. **357**(1), 241–278 (2006)
52. L. Fajstrup, M. Raussen, É. Goubault, E. Haucourt, Components of the fundamental category. Appl. Categ. Struct. **12**(1), 81–108 (2004)
53. L. Fajstrup, J. Rosický, A convenient category for directed homotopy. Theory Appl. Categ. **21**(1), 7–20 (2008)
54. D.R. Farkas, The Anick resolution. J. Pure Appl. Algebr. **79**, 271–279 (1992)
55. M.J. Fischer, N.A. Lynch, M.S. Paterson, Impossibility of distributed consensus with one faulty process. J. ACM (JACM) **32**(2), 374–382 (1985)
56. P. Gabriel, M. Zisman, *Calculus of Fractions and Homotopy Theory* (Springer, New York, 1967)
57. P. Gaucher, A model category for the homotopy theory of concurrency. Homol. Homotopy Appl. **5**(1), 549–599 (2003)
58. P. Gaucher, Towards a homotopy theory of process algebra. Homol. Homotopy Appl. **10**(1), 353–388 (2008)
59. P. Gaucher, Homotopy theory of labelled symmetric precubical sets. New York J. Math. **20**, 93–131 (2014)
60. P. Godefroid, Partial-order methods for the verification of concurrent systems - an approach to the state-explosion problem. Ph.D. thesis, Université de Liège (1995)
61. B. Goetz, J. Bloch, J. Bowbeer, D. Lea, D. Holmes, T. Peierls, *Java Concurrency in Practice* (Addison-Wesley Longman, Amsterdam, 2006)
62. U. Goltz, W. Reisig, The non-sequential behaviour of Petri nets. Inf. Control **57**(2), 125–147 (1983)
63. É. Goubault, The geometry of concurrency. Ph.D. thesis, École Normale Supérieure and École Polytechnique (1995)
64. É. Goubault, A semantic view on distributed computability and complexity, in *Proceedings of the 3rd Theory and Formal Methods Section Workshop* (Imperial College Press, 1996)
65. É. Goubault, Optimal implementation of wait-free binary relations, in *Proceedings of the 22nd CAAP* (Springer, New York, 1997)
66. É. Goubault, Some geometric perspectives in concurrency theory. Homol. Homotopy Appl. **5**, 95–136 (2003)
67. É. Goubault, E. Haucourt, A practical application of geometric semantics to static analysis of concurrent programs, *CONCUR 2005-Concurrency Theory* (Springer, New York, 2005), pp. 503–517
68. É. Goubault, E. Haucourt, Components of the fundamental category II. Appl. Categ. Struct. **15**(4), 387–414 (2007)
69. É. Goubault, E. Haucourt, S. Krishnan, Covering space theory for directed topology. Theory Appl. Categ. **22**(9), 252–268 (2009)
70. É. Goubault, E. Haucourt, S. Krishnan, Future path-components in directed topology. Electron. Notes Theor. Comput. Sci. **265**, 325–335 (2010)
71. É. Goubault, T. Heindel, S. Mimram, A geometric view of partial order reduction. Proc. Math. Found. Program. Semant. Electronic. Notes Theor. Comput. Sci. **298**, 179–195 (2013)
72. É. Goubault, T.P. Jensen, Homology of higher-dimensional automata, *Proceedings of CONCUR* (Springer, New York, 1992)

73. É. Goubault, S. Mimram, Formal relationships between geometrical and classical models for concurrency. Electron. Notes Theor. Comput. Sci. **283**(0):77–109 (2012) (Proceedings of the Workshop on Geometric and Topological Methods in Computer Science (GETCO))

74. É. Goubault, S. Mimram, C. Tasson, Iterated chromatic subdivisions are collapsible. Appl. Categ. Struct. **1**, 1–42 (2014)

75. É. Goubault, S. Mimram, C. Tasson, From geometric semantics to asynchronous computability, *Proceedings of DISC*, vol. 9363, LNCS (Springer, New York, 2015)

76. M. Grandis, Directed homotopy theory, I. The fundamental category. Cahiers de Topologie et Géométrie Différentielle Catégoriques **44**(3), 281–316 (2003)

77. M. Grandis, *Directed Algebraic Topology: Models of Non-Reversible Worlds*, vol. 13, New Mathematical Monographs (Cambridge University Press, Cambridge, 2009)

78. M. Gromov, *Hyperbolic Groups* (Springer, New York, 1987)

79. T.O. Group, Base specifications, POSIX.1-2008 (2013). http://pubs.opengroup.org/onlinepubs/9699919799/

80. J.R.J. Groves, Rewriting systems and homology of groups, in *Groups - Canberra 1989*, vol. 1456, Lecture Notes in Mathematics, ed. by L.G. Kovacs (Springer, New York, 1991), pp. 114–141

81. Y. Guiraud, P. Malbos, Polygraphs of finite derivation type (2014). arXiv preprint arXiv:1402.2587

82. J. Gunawardena, Homotopy and concurrency. Bull. EATCS **54**, 184–193 (1994)

83. F. Haglund, D.T. Wise, Special cube complexes. Geom. Funct. Anal. **17**(5), 1551–1620 (2008)

84. A. Hatcher, *Algebraic Topology* (Cambridge University Press, Cambridge, 2002)

85. E. Haucourt, Topologie Algébrique Dirigée et Concurrence. Ph.D. thesis, CEA LIST and Université Paris 7 (2005)

86. E. Haucourt, Categories of components and loop-free categories. Theory Appl. Categ. **16**(27), 736–770 (2006)

87. E. Haucourt, Streams, d-spaces and their fundamental categories. Electron. Notes Theor. Comput. Sci. **283**, 111–151 (2012)

88. M. Herlihy, D. Kozlov, S. Rajsbaum, *Distributed Computing Through Combinatorial Topology* (Elsevier, Amsterdam, 2014)

89. M. Herlihy, N. Shavit, The asynchronous computability theorem for t-resilient tasks, in *Proceedings of the Twenty-Fifth Annual ACM Symposium on Theory of Computing* (ACM, 1993), pp. 111–120

90. M. Herlihy, N. Shavit, The topological structure of asynchronous computability. J. ACM (JACM) **46**(6), 858–923 (1999)

91. M. Herlihy, J. Wing, Linearizability: a correctness condition for concurrent objects. ACM Trans. Program. Lang. Syst. **12**(3), 463–492 (1990)

92. C.A.R. Hoare, Monitors: an operating system structuring concept. Commun. ACM **17**(10), 549–557 (1974)

93. C.A.R. Hoare, Communicating sequential processes. Commun. ACM **21**(8), 666–677 (1978)

94. G.J. Holzmann, The model checker SPIN. IEEE Trans. Softw. Eng. **23**(5), 279–295 (1997)

95. J.F. Jardine, Cubical homotopy theory: a beginning. Technical report, Preprints of the Newton Institute, NI02030-NST (2002)

96. A. Joyal, M. Nielsen, G. Winskel, Bisimulation and open maps, *Proceedings of LICS'93* (ACM Press, New York, 1993)

97. T. Kahl, The homology graph of a higher dimensional automaton (2013). arXiv preprint arXiv:1307.7994

98. D.J. Kavvadias, E.C. Stavropoulos, Evaluation of an algorithm for the transversal hypergraph problem, *Algorithm Engineering* (Springer, New York, 1999), pp. 72–84

99. S.C. Kleene, *Representation of Events in Nerve Nets and Finite Automata*, Automata Studies (Princeton University Press, Princeton, 1956)

100. Y. Kobayashi, Complete rewriting systems and homology of monoid algebras. J. Pure Appl. Algebr. **65**, 263–275 (1990)

101. D. Kozen, Lower bounds for natural proof systems, *FOCS* (IEEE Computer Society, 1977), pp. 254–266
102. D. Kozlov, *Combinatorial Algebraic Topology*, vol. 21, Algorithms and Computation in Mathematics (Springer, New York, 2008)
103. D. Kozlov, Chromatic subdivision of a simplicial complex. Homol. Homotopy Appl. **14**(2), 197–209 (2012)
104. D.N. Kozlov, Topology of the view complex. Homol. Homotopy Appl. **17**(1), 307–319 (2015)
105. S. Krishnan, A convenient category of locally preordered spaces. Appl. Categ. Struct. **17**(5), 445–466 (2009)
106. S. Krishnan, Cubical approximation for directed topology I. Appl. Categ. Struct. **23**, 1–38 (2013)
107. Y. Lafont, F. Métayer, Polygraphic resolutions and homology of monoids. J. Pure Appl. Algebr. **213**(6), 947–968 (2009)
108. Y. Lafont, A. Prouté, Church-rooser property and homology of monoids. Math. Struct. Comput. Sci. **1**(03), 297–326 (1991)
109. L. Lamport, Proving the correctness of multiprocess programs. IEEE Trans. Softw. Eng. **3**(2), 125–143 (1977)
110. L. Lamport, How to make a multiprocessor computer that orrectly executes multiprocess programs. IEEE Trans. Comput. **28**(9), 690–691 (1979)
111. S.M. Lane, I. Moerdijk, *Sheaves in Geometry and Logic: A First Introduction to Topos Theory* (Springer, New York, 1992)
112. S. Lang, *Algebra*, 3rd edn., Graduate Texts in Mathematics (Springer, New York, 2002)
113. X. Leroy, D. Doligez, A. Frisch, J. Garrigue, D. Rémy, J. Vouillon, The OCaml system release 4.02 (2014)
114. J.J. Lévy, Réductions correctes et optimales dans le λ-calcul. Ph.D. thesis, Paris 7 University (1978)
115. D.R. Licata, G. Brunerie, A cubical approach to synthetic homotopy theory, in *30th Annual ACM/IEEE Symposium on Logic in Computer Science (LICS)* (2015), pp. 92–103
116. D.R. Licata, R. Harper, 2-dimensional directed type theory. Electron. Notes Theor. Comput. Sci. **276**, 263–289 (2011)
117. D.R. Licata, R. Harper, Canonicity for 2-dimensional type theory, in *Proceedings of the 39th ACM SIGPLAN-SIGACT Symposium on Principles of Programming Languages, POPL 2012*, Philadelphia, Pennsylvania, USA, 22–28 January 2012, pp. 337–348
118. W. Lipski, C.H. Papadimitriou, A fast algorithm for testing for safety and detecting deadlocks in locked transaction. ALGORITHMS: J. Algorithms **2**, 211–226 (1981)
119. N.A. Lynch, *Distributed Algorithms* (Morgan Kaufmann, San Francisco, 1996)
120. S. Mac Lane, *Categories for the Working Mathematician*, vol. 5, 2nd edn., Graduate Texts in Mathematics (Springer, New York, 1998)
121. A. Mazurkiewicz, Concurrent program schemes and their interpretations. Technical report DAIMI Rep. PB-78, Aarhus University (1977)
122. G.H. Mealy, A method to synthesizing sequential circuits. Bell Syst. Tech. J. **34**, 1045–1079 (1955)
123. R. Meshulam, M. Raussen, Homology of spaces of directed paths in Euclidean pattern spaces (2015). arXiv preprint arXiv:1512.05978
124. R. Milner, *Communication and Concurrency* (Prentice-Hall Inc, New York, 1989)
125. R. Milner, F. Moller, Unique decomposition of processes. Theor. Comput. Sci. **107**, 357–363 (1993)
126. S. Mimram, Sémantique des jeux asynchrones et réécriture 2-dimensionnelle. Ph.D. thesis, Université Paris-Diderot-Paris VII (2008)
127. A. Miné, Static analysis of run-time errors in embedded critical parallel C programs, in *Proceedings of the 20th European Symposium on Programming (ESOP'11)*, vol. 6602, Lecture Notes in Computer Science (LNCS) (Springer, New York, 2011), pp. 398–418
128. E.F. Moore, Gedanken-experiments on sequential machines. Ann. Math. Stud. **34**, 129–153 (1956)

129. L. Nachbin, *Topology and Order*, Mathematical Studies (Van Nostrand, Princeton, 1965)
130. T. Nakayama, J. Hashimoto, On a problem of G. Birkhoff. Proc. Am. Math. Soc. **1**, 141–142 (1950)
131. F.Z. Nardelli, Reasoning between programming languages and architectures, HDR memoir (2013)
132. M. Nielsen, G. Plotkin, G. Winskel, Petri nets, event structures and domains, part 1. Theor. Comput. Sci. **13**, 85–108 (1981)
133. M. Nielsen, V. Sassone, G. Winskel, *Relationships Between Models of Concurrency* (Springer, New York, 1994)
134. N. Ninin, Factorisation et représentation des régions cubiques. Ph.D. thesis, Université d'Orsay (2015). To be written
135. M. Nivat, Sur la synchronisation des processus. Rev. Techn. Thomson-CSF **11**, 899–919 (1979)
136. S.Y. Oudot, Persistence Theory: From Quiver Representations to Data Analysis. AMS Mathematical Surveys and Monographs. American Mathematical Society (2015). To appear
137. C.H. Papadimitriou, The serializability of concurrent database updates. J. ACM **26**(4), 631–653 (1979)
138. C.H. Papadimitriou, Concurrency control by locking. SIAM J. Comput. **12**(2), 215–226 (1983)
139. C.A. Petri, Communication with automata. Ph.D. thesis, Hamburg University (1962)
140. G. Plotkin, V. Pratt, Teams can see pomsets (preliminary version), in *Proceedings of the DIMACS Workshop on Partial Order Methods in Verification* (AMS Press Inc, 1997), pp. 117–128
141. V. Pratt, Modeling concurrency with geometry, in *Proceedings of the 18th ACM Symposium on Principles of Programming Languages* (ACM Press, 1991), pp. 311–322
142. F.P. Preparata, M.I. Shamos, *Computational Geometry: An Introduction*, 2nd edn., Texts and Monographs in Computer Science (Springer, New York, 1985)
143. M.O. Rabin, D. Scott, Finite automata and their decision problems. IBM J. Res. Dev. **3**, 114–125 (1959)
144. G. Ramalingam, Context-sensitive synchronization-sensitive analysis is undecidable. ACM Trans. Program. Lang. Syst. **22**(2), 416–430 (2000)
145. M. Raussen, Trace spaces in a pre-cubical complex. Topol. Appl. **156**(9), 1718–1728 (2009)
146. M. Raussen, Simplicial models for trace spaces. Algebr. Geom. Topol. **10**, 1683–1714 (2010)
147. M. Raussen, Execution spaces for simple higher dimensional automata. Appl. Algebr. Eng. Commun. Comput. **23**, 59–84 (2012)
148. M. Raussen, Simplicial models for trace spaces II: general higher-dimensional automata. Algebr. Geom. Topol. **12**(3), 1745–1765 (2012)
149. M. Raussen, K. Ziemiański, Homology of spaces of directed paths on Euclidean cubical complexes. J. Homotopy Relat. Struct. **9**(1), 67–84 (2014)
150. M.E. Saks, F. Zaharoglou, Wait-free k-set agreement is impossible: the topology of public knowledge, in *STOC* (1993), pp. 101–110
151. G. Segal, Classifying spaces and spectral sequences. Inst. Hautes Études Sci. Publ. Math. **34**, 105–112 (1968)
152. C. Sergio, L. Yuanxin, M. Andrea, J. Snoeyink, Testing homotopy for paths in the plane, in *Proceedings of the Eighteenth Annual Symposium on Computational Geometry* (2002)
153. J.P. Serre, Homologie singulière des espaces fibrés. Ph.D. thesis, Ecole Normale Supérieure (1951)
154. J. Sevcík, V. Vafeiadis, F.Z. Nardelli, S. Jagannathan, P. Sewell, Relaxed-memory concurrency and verified compilation, in *Proceedings of the 38th ACM SIGPLAN-SIGACT Symposium on Principles of Programming Languages, POPL 2011*, Austin, TX, USA, 26–28 January 2011, pp. 43–54
155. E. Soisalon-Soininen, D. Wood, An optimal algorithm for testing for safety and detecting deadlocks in locked transaction systems, in *Symposium on Principles of Database Systems (PODS'82)*, March 1985, pp. 108–116